Fundamental Theories of

C000256910

Volume 203

The international monograph series "Fundamental Theories of Physics" aims to stretch the boundaries of mainstream physics by clarifying and developing the theoretical and conceptual framework of physics and by applying it to a wide range of interdisciplinary scientific fields. Original contributions in well-established fields such as Quantum Physics, Relativity Theory, Cosmology, Quantum Field Theory, Statistical Mechanics and Nonlinear Dynamics are welcome. The series also provides a forum for non-conventional approaches to these fields. Publications should present new and promising ideas, with prospects for their further development, and carefully show how they connect to conventional views of the topic. Although the aim of this series is to go beyond established mainstream physics, a high profile and open-minded Editorial Board will evaluate all contributions carefully to ensure a high scientific standard.

More information about this series at http://www.springer.com/series/6001

Gregg Jaeger · David Simon ·
Alexander V. Sergienko · Daniel Greenberger ·
Anton Zeilinger
Editors

Quantum Arrangements

Contributions in Honor of Michael Horne

 Springer

Editors
Gregg Jaeger
Boston, MA, USA

David Simon
North Easton, MA, USA

Alexander V. Sergienko
Boston, MA, USA

Daniel Greenberger
New York, NY, USA

Anton Zeilinger
Vienna, Austria

ISSN 0168-1222   ISSN 2365-6425 (electronic)
Fundamental Theories of Physics
ISBN 978-3-030-77369-4   ISBN 978-3-030-77367-0 (eBook)
https://doi.org/10.1007/978-3-030-77367-0

This Springer imprint is published by the registered company Springer Nature Switzerland AG
The registered company address is: Gewerbestrasse 11, 6330 Cham, Switzerland

Introduction

Mike Horne left his home of Mississippi after completing his college degree in 1975 and arrived in Boston to study physics, as a graduate student at Boston University under the supervision of the physicist-philosopher Abner Shimony who had himself recently arrived from MIT. He spent the rest of his life in Boston, buying a house in the Dorchester neighborhood of the city, teaching for over 45 years at nearby Stonehill College, and working with collaborators at Boston University, MIT, and in Vienna. He and his wife Carole, who worked for decades at the Harvard Book Store, became local fixtures, spending much of their spare time frequenting local restaurants, independent bookstores and theaters, and clubs where Mike often himself played in jazz groups where *musical* arrangements came into play, complementing the quantum mechanical arrangements considered during his research life. Indeed, in addition to his intellectual achievements, Mike was well known for his friendly nature and his kindness to friends, colleagues, students, and anyone else he happened to meet. He often spent much of his day sitting in the hall of the Stonehill Physics Department or in the atrium of the science building simply talking to anyone who was there about politics, music, history, physics, or anything else that crossed his mind. His students would often spend hours mesmerized by his stories about physics and its history. Dinners at Mike and Carole's house would always be remembered fondly by all participants for the excellent food, music, conversation, and warm and informal atmosphere.

This volume begins with a chapter by Carole Horne entitled "Remembering Mike." There, Carole discusses her life with Mike and recounts some of his more interesting interactions with friends and colleagues, providing a window into his unique personality and giving the reader an idea of why he will be so badly missed by all who knew him; of the recollections of Mike by his friends she writes, "If I made a word cloud of what I heard these are the words and phrases that would be in huge type: kind, gentle, modest, unpretentious, generous, curious, smart, funny, enthusiastic, passionate, a great storyteller, an inspiring teacher, a creative physicist and researcher." Carole emphasizes how much Mike loved to collaborate in all aspects of life and how he, much like Richard Feynman, enjoyed the hands-on approach to things by working with their parts, especially building them.

After Carole's illumination of the man and the character and a broad history of his physics collaborations and what he saw as its close connection to teaching, the reader next encounters a chapter filled with Mike's own words, now focused mainly on his work as a research physicist in his interview with the late Joan Bromberg and stored the American Institute of Physics' oral history of quantum physics of the Niels Bohr Library and Archives, in which he recollects first meeting his collaborators who quickly also became his friends, among whom are two of us editors of this volume (Anton Zeilinger and Daniel Greenberger) and contributors to it (John Clauser and Carole Horne) as well as his experiences as a graduate student of Abner at Boston University and a post-doctoral worker under Clifford Shull at MIT. The discussion of the interview centers in particular on Mike's work on Bell's theorem and quantum entanglement, which he referred to as simply "Bell physics." Neutron and photon physics research of collaborating research groups in Boston and Vienna was the realm of investigation, always with an eye in the end toward practical experimental tests, of which the "Clauser–Horne–Shimony–Holt" theorem became a towering mathematical result of 20th Century physics. In addition to describing how such results of his research came about, Mike also explains how he came to take his place as a professor of physics at Stonehill College, where he introduced a physics major for students pursuing bachelor's degrees and how the college was a perfect place for him to pursue his research.

One of Mike's major contributions to the foundations of modern physics, where physics meets philosophy, is the development, along with John Clauser, of the idea of "Clauser–Horne (CH) Local Realism." Roughly speaking, this is the idea that, in Mike's words, the world is made of "real stuff" that exists independent of any measurements and that obeys the causal rules imposed by special relativity. The "stuff" of CH Local Realism is a more precise version of what John Bell referred to as "beables" and Einstein, Podolsky, and Rosen called "elements of reality". Forty years of experiments testing the Clauser–Horne–Shimony–Holt (CHSH) inequality and other related Bell-type inequalities have shown unambiguously that CH Local Realism is violated. In his contribution to this volume, Clauser studies two formulations of quantum mechanics: The lab-space formulation, often used to study single-particle quantum mechanics, and the configuration-space version that is frequently employed in many-body quantum mechanics. He shows that these two formulations are inequivalent. This is in contradiction to the conclusion of Max Born, one of the earliest advocates of the lab-space formulation, who proclaimed them to be equivalent. Clauser concludes that lab-space formulations are untenable, as they do not correctly predict the expected violation of CH Local Realism.

Although the CHSH inequality has generally been used to tests the predictions of quantum theory with entangled photon states, another major component of Mike Horne's career involved the analysis of neutron diffraction experiments utilizing crystals. As always, Mike's interest was to use these experiments to shine light on fundamental aspects of quantum mechanics itself. Much of this work was done in collaboration with (future Nobel laureate) Shull and his students and other postdocs at MIT. Using the dynamical theory of diffraction, they looked at the effects of spin-orbit coupling in neutron–nucleus scattering and found a resonant enhancement of the spin-orbit scattering when an external magnetic field causes the Larmor precession distance to coincide with the Pendellösung distance in the crystal. Pendellösung is the periodic flow of energy between forward and backward Bragg scattering amplitudes in a crystal. Mike seems to have been the first to notice this enhancement, now known as the Neutron Spin-Pendellösung Resonance (NSPR) effect, and he derived the corresponding differential equations that describe the effect. Kenneth Finkelstein first discusses the theoretical description of NSPR and then describes an experiment that demonstrates the resonant enhancement effect and yields a precise value for the ratio of spin-orbit scattering to nuclear scattering in silicon. Those experimental results reveal a discrepancy with theoretical calculations. The physical principles underlying the required theoretical correction are then explored.

It has been pointed out by Mike's long-term colleague, friend, collaborator, and co-editor of our volume, Daniel "Danny" Greenberger, that Mike was always eager to reach the bottom of every physical effect and never be satisfied until he fully understood all the specific details and implications of them; Mike was always happy to share this understanding that often helped dispel some misconceptions in interpretation of quantum processes. One example of his scientific approach to complex physics topics is described in the next contribution here by Danny that deals with understanding multiple conceptual questions surrounding the Aharonov–Bohm effect. This piece offers an extremely clear and physically concise introduction to one of the effects most discussed over the last several decades in physics—the description is rigorous but sufficiently transparent to be understood by non-experts in that particular field as well as by philosophers of science.

The main issue discussed by Danny Greenberger in his chapter is how one could understand the non-classical nature of the Aharonov–Bohm effect and explain the modulation of electron's phases while there is no real magnetic field present in a space outside the solenoid. He points out a very original contribution of Mike Horne to the study of the AB Effect that offers an interesting interpretation dealing with the need to set up initial experimental conditions by using real fields and forces prior to conducting the final experiment with stationary magnetic fields inside and outside the solenoid. This concept—that of taking into consideration all energy-related parameters contributing to the quantum system under a considered preparation—has significant implications and helps to understand the nature of many conceptual paradoxes surrounding quantum mechanical effects.

The general notion and specific features of quantum entanglement—the implications of which Mike and his collaborators have pioneered—have been at the heart of many quantum mechanical and, more recently, quantum information processing effects that have occupied the minds of researchers and philosophers now for a half-century. The multiple manners of quantifying entanglement that can be found in specific physical systems, as well as elaborate schemes for their observation that have been developed have been impacted by their work. These scientific and intellectual challenges were the perennial subject of Mike's research, and the fruits of his unique approach to doing physics left significant impact on this field by actively participating in the construction of two major Bell's inequalities that have been widely used in the field.

Mike Horne's enthusiasm and dedication to resolving such complex issues relating to entanglement are also highlighted in personal recollection by Anton Zeilinger, Marek Zukowski, and Caslav Brukner which follows. The use of information as an indicator and the connection between quantum systems or between parts of a single complex quantum system is one of the most intuitive and informative approaches to understanding entanglement. Many specific details of this useful and informative concept are discussed in great detail in this chapter by these three major experts in the field. One of the central outcomes of this contribution is the demonstration of the universality of their approach which is exhibited through the equivalence between two entanglement evaluation approaches based on the information theoretic and Bell's inequalities formalisms.

Herbert Bernstein's contribution to this demonstrates Mike Horne's interest in photon physics and in collaboration, particularly his joint work with Cliff Shull, Anton Zeilinger, and Danny Greenberger. They all sought to answer the question "*Why* the quantum?" by another great of the field, John Archibald Wheeler. Bernstein's contribution is an experiment aimed at providing evidence that each individual particle can be described by a corresponding state, that is, wavefunction that would distinguish its state from a statistical state. The experiment invokes the technique of heralded single-photon detection. A key technique of the experiment is to create "signal" photons by parametric down-conversion in pairs, one of which, the idler, heralds the presence of the corresponding signal photon to the experimental apparatus. The signal photons are prepared in different states so that no individual-system state appears is prepared more than once in the entire set of preparations. The set of individual single-particle measurements resulting is then to be compared with the predications of standard quantum mechanics. The observable property at the center of the experiment is that of the polarization degree of freedom: For the linear polarization preparations, this treatment demonstrates that the cosine-squared Law of Malus is confirmed by preparing the individual systems in all a full set of differing orientations and measuring the passage of the systems through a polarization analyzer of fixed orientation.

The work of Mike Horne was always aimed at either a better understanding the fundamental elements of quantum mechanics or the surprising implications of those fundamental elements. One of the most unique aspects of the foundations of quantum mechanics as physics has come to grips with them has been the physics of measurement. Don Howard's contribution to this volume considers the various means by which the measurement problem of quantum mechanics, the problem that the modeling of measurement by the linear state evolution of the Schrödinger equation alone, predicts indefinite outcomes of measurement when the measuring device is part of the system studied. In particular, Howard suggests taking this physical law seriously to see whether the appearance of definite measurement outcomes could be an illusion brought about by quantum state decoherence. He notes that the name applied to the study of this sort of behavior, *decoherence theory*, is a confusing one in that it suggests that so-called state collapse is what takes place under this phenomenon whereas quite the opposite is true: Coherence is retained in composite systems but relates to what can be measured in a far subtler way than is present, for example, when one has a simple coherent wave-packet as the spatial representation of a non-composite system. Howard explores question of what sort of "observational indistinguishability" of states corresponding to different measurement outcomes amounts to, which he identifies as a puzzle to be solved. This analysis begins with the reconsideration of the notion of complementarity in the thought of Niels Bohr and the understanding of entanglement before Erwin Schrödinger's naming of it first lurked in the minds of Bohr, Albert Einstein, Boris Podolsky, and Nathaniel Rosen who together made the notion of physical realism one that has been directly pursued for nearly a century.

To set the stage for understanding this notion both physically and philosophically, Howard lays out the history of thinking about entanglement in the late 1920s and early 1930s and the evidence of its commonplace consideration. He notes that Einstein considered in relation to the statistics of multi-particle systems, Bose vs. Boltzmann statistics, in particular. Other examples, in writings of Hermann Weyl and Wolfgang Pauli are also noted. Howard then takes up the question of the relation between entanglement and complementarity, that entanglement entails complementarity, at least in Bohr's own sense of the notion. Another important aspect of the consideration of the foundations of quantum theory is the distinction between the classical and the non-classical, which is connected in the physics of that era with the appearance of measurement outcomes, most often understood capable of appearing in measuring systems that, at least when alone, are well modeled by classical physics. Howard carefully considers those characteristics commonly attributed to possible measuring instruments, such as mass, size, and number of degrees of freedom, and critically so. This, together with Bohr's notion of a "phenomenon," is then assembled into a unified picture of Bohr's conception of what takes place in the measurement of a quantum system.

Such a picture of physics allowed Bohr to advance the thesis that quantum and classical physical descriptions can be essentially equivalent. Howard illustrates this via the consideration of context-dependent mixtures, to be distinguished from improper mixtures represented by reduced density matrices. They are, rather, joint density matrices representing proper mixtures that "can be interpreted as if they represented, *with respect to the degrees of freedom measurable in the stipulated context and only those degrees of freedom*, mutually independent systems." In that respect, they correspond to what Bohr regarded as classical descriptions.

All of the above enables Howard to present a dissolution of the measurement problem along the above lines: "Applying linear, Schrödinger dynamics to the system-instrument-environment interaction drives the joint, system-instrument-environment state into an entangled, pure, joint state that is observationally indistinguishable from the relevant context-dependent mixture picked out by the measurement context because the pure, joint state and that mixture over joint eigenstates picked out by the measurement context give exactly the same statistical predictions for all observables measurable in that context." What is offered is a mathematical equivalence, not simply some imprecise suggestion of one. Thus, finally, he argues that environment-induced decoherence "was, all along, the real point toward which Bohr was gesturing with the doctrines of complementarity and classical concepts."

Another issue to emerge in the history of quantum physics that owes much to Bohr is complementarity. And, not surprisingly, Mike Horne and his collaborators (including two of the editors of this volume, Gregg Jaeger, Mike's former thesis-advisor Abner Shimony, and Anton Zeilinger) also probed the complementarity between single-particle interference, for which some offer more classical explanations, and quantum entangled-state *multi-particle interference*, which has no adequate classical explanation. Quantum interference had been at the focus of extensive theoretical and experimental investigation for years, but these workers were expanding this consideration to systems of their complementarity as well as the consideration of the joint interference of more than two quantum particles. The nature of such higher-order interferences and associated complementarities remains an area of active research. Mike's scientific curiosity and interest in solving challenging problems homed in on this topic quickly; his vision of the problem and the essence of this exciting quantum physical challenge are presented with careful attention to detail by Christoph Daniel and Gregor Weihs in their contribution here. While previous work in this area has focused on the effect of local state transformations on multi-particle states, the contribution of Daniel and Weihs examines the effect of global transformations, showing that the complementarity rules derived by Mike Horne and collaborators naturally extend to this case when the role of quantum entanglement is properly accounted for. They also note, in light of their new results as well as Mike's own, that his personal goal of a three-body complementarity relation involving single particle, two particles, and three particles that any such three-body complementarity relation would preferably be addressed in terms of entanglement between the constituents rather than ordinary interference visibilities.

Another of the significant theoretical results Mike Horne produced in his collaborations is the Greenberger–Horne–Zeilinger (GHZ) theorem, which involves the consideration of quantum states of a four-qubit system. They already mentioned that the same results would hold for three-qubit systems. The broader context is that of values (plus or minus one) that can be assigned by any putative non-contextual hidden-variables model for quantum mechanics. Later, David Mermin considered such a set of observables in the context of the three-qubit system. The GHZ theorem is based on the consideration of quantum state vectors—now widely used and called "GHZ states"—in which the values of the quantities involved are correlated in the particular ways the significance of which was first pointed out by this trio. In the final contribution to this volume, Mordecai Waegell and P. K. Aravind explore logical relations between propositions in quantum theory by relating the original result of GHZ to David Mermin's different proof of their result, one of the Kochen–Specker theorems based on the ten GHZ observables. These quantities can be arranged elegantly along the edges of a pentagram with values assigned to the observables being plus or minus one that can be viewed as an edge-coloring problem.

Mermin showed how the correlation requirements imposed on these values preclude a logically consistent assignment of values "at once" to all these quantities. Waegell and Aravind demonstrate how this proof can be transmuted into a different one for the same result, that of Kernaghan and Peres, based on the eigenstates of those observables. The Kernaghan–Peres result was arrived at by returning again to the consideration of quantum state vectors, with the price, again, of a slight loss of simplicity. This shift of approach is possible because these proofs are essentially "parity proofs," that is, ones based on an even–odd contradiction. Waegell and Aravind show that the transformation of perspectives, which they called a "looking glass" relation, has proven valuable in understanding other aspects of quantum theory in the last thirty years.

This volume in honor of Mike Horne offers not only many fond recollections of Mike as a person and documents important aspects of his work as a scientist, but also gathers some of the most recent work that shows that his impact on his collaborators and physics at large continues.

Contents

Chapter 1
Remembering Mike

Carole Horne

When Mike died I heard from many, many people who knew him—friends, fellow physicists, fellow faculty members, students, musicians – and I've heard from many more since. I've gotten used to meeting people—neighbors, people in stores and restaurants, electricians, and mail carriers—who have stories about the ways in which Mike affected their lives. If I made a word cloud of what I heard, these are the words and phrases that would be in huge type: kind, gentle, modest, unpretentious, generous, curious, smart, funny, enthusiastic, passionate, a great storyteller, an inspiring teacher, a creative physicist and researcher. Readers who knew him will recognize Mike in these recollections, I think.

I'd like to write about those qualities that, for me, get to the heart of who Mike was. First and foremost, he was the most joyful person I've ever known. He was full of wonder at existence, excited and fascinated by the world and what we can understand of it. Many mornings, lying in bed, he'd wake up, wake me up, and as we shook off our sleepiness would say with a little smile, "Ain't life grand?" And he often said, apropos of nothing in particular, "Isn't it great to be alive." He actually did say those things. He was quite aware of, and outraged by, all the things wrong with the world, but nevertheless he felt that the two of us were extraordinarily lucky. I've never known anybody else like that. Many people who wrote to me said things like "his love of life was contagious, his joy infectious." And it was. He threw himself wholeheartedly into the many things he loved.

And then there was Mike's impulse to share everything he loved or discovered. It was most obvious in his physics. He loved to collaborate. From his early grad school days in the basement office he shared with ten or so other students, he was happiest when they all talked about what they were working on, critiquing each other's ideas, adding to each other's efforts, cheering each other on when someone

C. Horne (✉)
Harvard Book Store, Cambridge, MA 02138, USA
e-mail: chorne@harvard.com

© Springer Nature Switzerland AG 2021
G. Jaeger et al. (eds.), *Quantum Arrangements*, Fundamental Theories of Physics 203,
https://doi.org/10.1007/978-3-030-77367-0_1

had a breakthrough. His earliest successes as a physicist were in collaboration with his dissertation advisor at Boston University, Abner Shimony, and John Clauser, and Richard Holt. He and Abner continued to collaborate long after Mike received his doctorate. Later, his collaborations with Danny Greenberger, Anton Zeilinger, Cliff Shull and others—which started in Cliff Shull's lab at MIT—lasted his whole career.

But the impulse to share and collaborate went beyond his physics. More than many couples, we shared most everything we did—music, cooking, traveling, hosting people at our house—and told each other about everything we encountered that excited us. Mike shared things with everyone. When he discovered a new movie, especially a classic movie from the 40s or 50s, he tried to convince everybody to watch it. When we successfully tried a new dish, he'd spread the recipe around. When he read something that he was especially impressed by, he'd make copies to carry around and pass out. If he discovered a new jazz musician or recording, there were any number of people who would hear about it. He played music with people as often as he could, and after all, what is jazz improvisation but a deep collaboration? And something new in physics? Everybody heard about it. When gravitational waves were detected at LIGO, he didn't stop telling people for weeks.

About his modesty, which virtually everyone mentions. More than most people, I think, Mike didn't worry much about what other people thought of him. In his career he wasn't ambitious in the usual sense; he simply wanted to understand things, and was happy and proud when he'd figured out something that he thought was important. Although he was recognized as a special teacher, he didn't think of his teaching as a way to advancement, but as a way to share with students a new view of the world. He didn't do things for the acclaim, although he appreciated it (in his low-key way) when it came. He had a very balanced and unassuming sense of himself, happy and confident, but not overly impressed with himself either. A relative of mine, who thought she'd be intimidated by someone with a Ph.D., once said she liked being around Mike because "he's just a regular person." I suspect this modesty was why children and young people were so drawn to him. A nephew wrote to me about Mike being the first adult who didn't talk down to him, and who listened to him seriously. He said he tries to do that with his own students now that he's a teacher.

Mike's attitude toward physics was unusual. He was passionate about it, but he came to it from books he read as a junior in high school. NSF had commissioned a series of science books, published by Doubleday Anchor, which included biographies of great physicists and stories about foundational discoveries in science. He always said he was more affected by the sense of what it meant to do physics—the excitement and human drama of the endeavor—than he was by the textbooks he studied. He once wrote "the physicist, when asked 'what is physics anyway, and why do you do it?' will talk about beauty, elegance, mystery, excitement, reality, humility, goose bumps, and tears." He was interested in the history and philosophy of science, and applied only to Boston University for grad school because he could study both physics and its history. In 1968, when Abner Shimony came to BU from MIT, he became Mike's dissertation advisor. What a lucky happenstance! Being both a physicist and a philosopher, Abner's interests were exactly right for Mike. At a time when mainstream physics was not much interested in foundations of physics, Mike was fascinated. It didn't matter to him—in fact, I doubt he ever thought about it—that his dissertation topic might not be a smart career move. Eventually though, the work he and Abner did helped to bring about a wide appreciation of nonlocality and entanglement as physics worth studying.

When he came home in 1976 from an early conference organized by John Bell in Erice, called Experimental Quantum Mechanics, I remember Mike's excitement telling me about having met this young Austrian physicist, Anton Zeilinger. They had apparently hit it off, staying up very late in the evenings, drinking wine and talking about foundations of quantum mechanics. Mike felt that he'd met someone who shared some of his thinking about physics, and he was thrilled when Anton came for the year as a postdoc to Cliff Shull's lab where Mike was on sabbatical from Stonehill College as a visiting scientist. Danny Greenberger first met Mike and Anton at a conference in Grenoble in 1978. They hit it off wonderfully too, and Danny started coming up from New York to Cliff's lab at MIT. The GHZ collaboration went on for 40 years. There were many conferences, including in Japan, Vienna, and New York (where I was a tag-along and got to experience the physics world Mike lived in, which became an important part of my life too). Perhaps the best conference, though, was in Amherst in April of 1990. There were no prepared talks, no schedule, no proceedings, just conversations. Mike thought it was the perfect idea for a meeting. There were a small number of people attending, including John and Mary Bell. Danny, Mike, and Anton talked about the GHZ theorem, which was new, and they were excited by Bell's interest. Sadly, Bell died unexpectedly that October.

Mike strongly believed that simplicity was the essence of physics; it's one of the things that drew him to the field. He started every semester by telling his students "if it's complicated it's not physics". Mike believed that teaching physics made him think about physics in simpler, clearer ways. He thought that if you couldn't explain something to someone with no physics background, without watering it down, you didn't really understand it yourself. It didn't matter whether he was talking to students or to fellow physicists, he wanted simple, elegant explanations. He was an unusually visual person, and I think that helped him see things more simply. A funny aside—there's a famous photo that has a cow in it, but the way it's taken you see an abstract photo, and don't see the cow. When I showed it to him and asked what he saw, he said "you mean besides the cow?" As a physics colleague related to me, "topological things seemed so natural coming from him. He was always explaining things in simple, uncluttered ways to all of us."

Mike was a dedicated teacher. Because not every student found physics simple to learn, he spent his time at the college not in his office but in the Atrium, available for any student to come get extra help. He often talked with his students about his love of physics, and his view that questions arise in physics not for practical reasons but "only if you want to understand." Just as he thought that teaching physics made him a better physicist, he thought that doing physics was essential to being a good teacher.

Doing physics and teaching physics were inextricably connected for Mike. For example, from his dissertation topic to the end of his career, although he was a theorist, he straddled the border, trying to devise doable experimental tests of existing theory.

He sometimes called this "quantum archeology," a phrase that Abner loved. His involvement with experiments transferred to his teaching. When he was beginning his teaching career, he wanted to develop a lab course in which students measured the fundamental constants using the original techniques. Unable to buy commercial versions of all the equipment, some theorists might have abandoned the idea. Because he thought exposure to these original experiments was so important, he learned how and built most of it himself.

Education, Mike thought, especially science education, could change your life. When Stonehill proposed dropping the general studies requirement from two courses to one, he wrote an argument against the change; he titled it (after a John McPhee book) "A Sense of Where You Are," and he spread it around campus. He told a story about going to college and eagerly signing up for three classes—history of science, astronomy, and physical anthropology. He said those courses gave him a new way of thinking about the world and our place in it. A few years ago he wrote, "To this day when I step outside in the morning, my inner voice often says something like 'I'm a recent hominid and I can do physics.' The tone of that inner voice is that of the Truman Capote character in the film *To Kill a Mockingbird*, who introduced himself with 'I'm seven and I can read'." That's what he wanted for his students.

Mike was thoroughly captivated by physics, especially the foundations. He thought that doing physics was a creative endeavor, in the same way that art and music are creative. He completely rejected the idea of physics as a dry, boring, difficult subject. He was so involved in whatever physics he was currently working on that when an idea came to him, he'd interrupt what he was doing to capture it. A friend reminded me of seeing Mike at a party, sitting in the corner, writing something down in a notebook. At the college, when he wasn't talking to students, he was doing physics with the pencil and notebook he always had with him.

Mike brought that kind of passion to his other loves: jazz, building things, politics, movies, cooking. A recent book, *The Quantum Dissidents: Rebuilding the Foundations of Quantum Mechanics (1950–1990)*, by Olival Freire, Jr., suggests that there may be a connection between the political activism of the mid-to-late twentieth century and the concurrent questioning of the foundations of physics. It's possible. For as long as I knew him, Mike had a deep belief in progressive politics. He talked about how his dad, born in 1896 in rural Mississippi, ran in the mid 30s for county school superintendent, and having won, promptly built a school for Black kids, who had not had a school until then. Equally promptly, burning crosses appeared in the front yard, and his dad wasn't re-elected. Although Mike wasn't born for another decade, he was profoundly affected by that story. He was a college sophomore and I was a freshman in 1962, when we met at the University of Mississippi the fall that James Meredith enrolled and became the first African American student to attend the university. Meredith was accompanied by U.S. Marshals and his arrival was met with several thousand people rioting, driving around the campus waving Confederate flags, shooting guns, burning cars. Two men were killed and National Guard troops were sent and remained on campus for the school year. This experience solidified for both of us our commitment to civil rights, to racial justice and equality, and to progressive politics. When Mike graduated in 1965, we married and moved to Boston for Mike to start graduate school. We never returned to live in the south.

Lots of people have talked about Mike's love of jazz, both playing it and listening to it. He thought jazz, like physics, could change your life, and as with physics, he wanted to tell everyone about it. As with physics, his gift was being able to explain musical things simply to people untrained in music: time, space, playing ahead of or behind the beat, minor keys, the way musicians listen to each other, they way they "talk" with their instruments. He could make you hear with bigger ears, make you understand in a deeper way. One of the best descriptions I have is from a relative who once picked up Mike's upright bass and tried to play along with the record that was on. At one point a series of notes sounded right, and Mike called from another room, "You got it!". "No Mike," he later wrote, "I didn't get it, you gave it to me."

Building things was another love of Mike's. From the oversized dining room table he built the first year we bought our house, using tools from the shop in the BU physics department, to bookcases all over the house, to the enormous project of rebuilding the carriage barn behind our house, there was always a building project underway. The carriage barn, built in 1874, had been moved from its original location to nearer the house, and had fallen off the badly-built new foundation. Restoring it was a project that took Mike and my brother, Larry, five summers, and entailed lifting the whole structure on specially made steel posts that he designed and had built, and building a new foundation. He was enormously proud of it when it was done. In 2015 we threw a party in the barn, complete with a whole roasted pig, to celebrate our 50th wedding anniversary and 140 years of the barn.

At its heart, the story of Mike's life is a love story. Boy falls in love with music at a young age. Boy meets girl, falls in love and marries. Boy discovers physics and teaching, falls in love and spends his life doing them, has a life rich with loving friendships. Like every human story, there's a sad ending, although Mike would disagree with that description. But like some human stories there's a lasting impact. Mike's enthusiasm for all the things he loved was boundless. As a colleague wrote, "maybe that was the key to his rich and multifaceted life: he was always up for having second helpings; he was on for another set." And he was always teaching—often by example—mostly how to live (Figs. 1.1, 1.2, 1.3 and 1.4).

A note on the players

There are so many people who made Mike's career a joy. I've talked about Abner Shimony, Mike's dissertation advisor, who grew up in Memphis and became a life-long collaborator. Mike's dissertation topic was completed simultaneously by John Clauser at Columbia, and Clauser agreed to collaborate on the publication with Mike, Abner, and Dick Holt. Holt was the graduate student at Harvard who planned to do the proposed experiment, the experimental test of the CHSH-Bell's theorem predictions. Mike and John went on to collaborate on several papers, most notably leading to the 1974 Clauser–Horne (CH) inequality.

When Mike decided to apply only to BU for grad school, it was because he knew that the physics department chair, Robert Cohen, was a co-founder of the Center for the Philosophy and History of Science, and its Boston Colloquium for Philosophy of

Fig. 1.1 Abner Shimony, Mike, and Anton Zeilinger

Science. The Center was founded by Bob and Marx Wartofsky in 1960 as an offshoot of the Institute for the Unity of Science, which was itself the American transplant of the historic Vienna Circle. So things came full circle in Mike's physics life, with connections to Vienna. Bob Cohen was important in two other ways: he helped, with Chuck Willis, convince Abner to come from MIT to BU, and he told Mike about an available job at Stonehill College.

In 1970, Chet Raymo, another southerner, and an accomplished naturalist and writer, hired Mike to teach the physics classes at Stonehill College, the perfect job for him: a small college where there was no pressure to publish for the sake of advancement, near Boston where he could work with other physicists. Mike always gave Chet credit for teaching him many things about good teaching.

Mike showed up unannounced at MIT in Cliff Shull's lab in 1976 to talk about a possible experiment for Shull's new neutron interferometer, and after talking a while

Fig. 1.2 Danny Greenberger, Mike, and Anton Zeilinger (GHZ)

Fig. 1.3 Group photo in Shull's lab (circa 1982?)—left to right, John Arthur, Anton Zeilinger, Cliff Shull, Mike Horne, Danny Greenberger, Ken Finkelstein, and Tony Klein

asked "Can I play?" Cliff gestured to an empty desk and said "Why don't you sit
there." Mike worked in the lab in all his free time for almost fifteen years. Danny
Greenberger started coming to the lab in 1978, and Herb Bernstein from Hampshire
College was in the lab as a visitor too. In Cliff's lab, Mike worked with a number
of students: Ken Finkelstein, Don Atwood, John Arthur, Steve Collins, and Dan
Gilden among them. Joe Callerame was a postdoc in the lab, working on building
the interferometer. Tony Klein from Melbourne, Australia was also there on several
extended visits.

After Shull retired in 1986, Mike, Danny, Herb, and Anton applied for NSF
grants that continued for about four years. Cliff continued to visit and to look over
the shoulders of students doing experiments in what he called the "remnants of my
old research laboratory."

After that, Mike made many trips to Austria to Anton's lab, where he met and
worked with a number of students, postdocs, and physicists over the years, most
notably Marek Zukowski from the University of Gdansk. He loved working with the
students, and always returned with a new idea to think about.

There were physicists that Mike had many conversations with, especially Helmut
Rauch, Anton's advisor in Vienna, Sam Werner at the University of Missouri, and
Yanhue Shih at Maryland. He also at one point started visiting David Pritchard's lab
at MIT.

Mike collaborated at various times at Boston University with Alexander Sergienko, Gregg Jaeger, who was also a PhD student of Abner Shimony's, and David Simon, who later joined the Stonehill Physics Department.

For many years at Stonehill, Chet Raymo taught courses for non-science students. His courses "The Earth" and "The Universe" were extremely popular. Mike taught the physics courses needed by chemistry, biology, math, and pre-med students. After Chet retired, the College decided that it needed a physics department, and Mike set about finding faculty. Alessandro Massarotti had come to Stonehill to replace Chet, and after that, they hired David Simon and Mevan Gunawardena. When Mevan returned to Sri Lanka, the department hired Ruby Gu. Mike was very pleased that the college had been able to attract such terrific scholars and teachers.

Chapter 2
American Institute of Physics Oral History: Michael Horne

Michael Horne

Interviewed by: Joan Bromberg, Department of History of Science and Technology at the Johns Hopkins University.[1]
    Location: Dorchester, MA Interview date: Thursday, 12 September 2002.

Horne: And then he kept coming back. You know, he'd come back for six months, stay, and he has made many, many visits to MIT over the late seventies and all through the eighties.

Bromberg: Oh. So he was —

Horne: But he was here.

Bromberg: I had better tell the tape, because I just turned it on. We are talking about Anton Zeilinger, whom Professor Horne met Erice in 1976.

Horne: That's right. And by that time I had just started working with Cliff [Clifford] Shull at the MIT reactor where he was doing the new work then on neutron interference. There were several groups that had started that before he had. Sam Werner [spelling?] had a group in Missouri and they had built an interferometry. You know, everybody was using the same techniques in these first interferometers — silicon crystals and thermal neutrons. And they were just essentially doing the double-split experiment of the neutrons. And Abner [Shimony] and I saw that, saw those first papers, the one from Missouri. The one from Missouri by Werner, their experiment was very nice. They turned the interferometer so that the effect of gravity could be seen, so they could actually see

[1] Note: This interview is reproduced from the online archive of the AIP Oral Histories (with permission); https://www.aip.org/history-programs/niels-bohr-library/oral-histories/34331.

M. Horne (✉)
Easton, MA 02357, USA
e-mail: nbl@aip.org

© Springer Nature Switzerland AG 2021   11
G. Jaeger et al. (eds.), *Quantum Arrangements*, Fundamental Theories of Physics 203,
https://doi.org/10.1007/978-3-030-77367-0_2

the phase shift of the radiation due to turning the interferometer in the gravity field. The other group that had started this business was Rauch and his student Zeilinger and others, and they were in Vienna at a reactor. And one of the first experiments they did was a demonstration of the effect if you turn a neutron around once using magnetic fields, if you rotate it once. Then it's not really like it was before you turned it. It actually introduces a negative sign in its quantum state, which with an interferometer is visible, so they could see this negative. If you turn it another time — so if you turn it twice — then it's back to its original condition. So those are the two experiments that caught our attention.

Bromberg: What made you —? I mean you were working up until this point on Bell's theorem.

Horne: Well, I thought Bell's theorem was finished. Right? That is the original work with Abner for my thesis, in collaboration with Clauser, and then, tidying up some questions we had with Clauser in that '74 paper, was trying to improve the argument a bit. And then by then the experiments were done. You know, Clauser had done the experiment in California, and the quantum mechanics was beautifully confirmed. The Harvard experiment that gave a disconfirmation of quantum mechanics. Everyone knew there was something wrong with it, they just didn't know what. So I figured we were through. In 1975 I was looking for some more fun things to do with quantum mechanics, and I saw these neutron interferometer papers coming out from the States and from (Proposed neutron interferometer observation of the sign change of a spin or due to 2 precession) Austria. Abner and I talked about them together and we even wrote a paper proposing that this turn a neutron around once and get a negative and then turn it twice and then it would be back to positive. We wrote a paper proposing that this experiment be done, not knowing that they were just finishing doing it in Vienna. It was a wonderful toy to see something as basic as interference which is the heart of quantum mechanics be doable now with something as massive as a neutron. We just thought, "This is something that is going to be fun to play with for a number of years." So that's why I made the transition. I thought the Bell physics was essentially done, the experiment had been done, the elements of reality don't seem to be possible because the quantum mechanics was confirmed. So I thought the Bell physics was essentially a finished package. So I saw these one-particle interferometers and thought I would like to play with them. And so I heard from someone — I don't know who told me that there might be soon one of these same instruments running at MIT because one of the founding fathers of neutron diffraction was a professor at MIT, Cliff Shull. So I don't remember who told me about him. He hadn't done any interference yet, but as it turned out he was working on it trying to get one running. So I went over there sometime in '75, I'm not sure when. I went over and walked into the reactor where his office was and he was there with a couple of his grad students, and I introduced myself and told him about my background in Bell physics and my work with Abner and my excitement about these new one-particle neutron interferometers, and I essentially asked him, "Can I play?" And he says, "Sure. Just take that desk." So I just started going over there every Tuesday, which was my day off at Stonehill, and weekends and summers and Christmas. That's how I started.

Bromberg: So you were really set up to meet Zeilinger, to meet someone from the Rauch group, I mean that was —

Horne: Well, then what happened was, completely separate from Shull — And by the way when I went over there he had a student who was Joe Callerame who was currently trying — they had already cut the crystal and they had a new neutron beam and they were trying to make it show interference, and during the first few months I was hanging around there it started working. I was there the night that — what Joe did was put pieces of aluminum foil in one of the beams. He put pieces of aluminum foil and that would phase shift the radiation in that beam relative to the other. So the aluminum foil was playing the role of glass plates, you know in optical interference. We wrote a paper. And he put in one sheet of Reynolds Wrap and then he put in another sheet of Reynolds Wrap and watched the counts go down, indicating the phase shift was happening. And then he kept putting in more sheets and then it came back up, right? So there was interference. So we called Shull at his home in Lexington and he drove in. It was late one night. We went out and had a drink because the interferometer was working. So this was the third one. You know, there was the one in Missouri and then there was the one in — [Vienna].

Bromberg: I see.

Horne: But at this time there was no connection with the Vienna group or with Zeilinger. That connection happened because of a little carryover of the two-particle stuff. There was a meeting and there was going to be a conference in Sicily in '76. I was already hanging out with Shull and the neutron interferometer people. There was going to be a conference on Bell physics. I didn't have anything new to present on the Bell physics, but you know it was something I had worked on and so I decided to go with Abner to the Sicily meeting. We also went with Frank Pipkin, who was the professor at MIT who had done the [Bell Theorem] experiment. So the three of us went together to Sicily to that meeting, and it was at that meeting that I ran into this person about my age — thirty-ish. Zeilinger had come to that meeting and his plan was to talk about their neutron experiments even though it wasn't directly connected to two-particle quantum mechanics or to Bell stuff. It was so basic and fundamental he thought that would be a good place to go talk about it.

Bromberg: So he was essentially also motivated to do physics that had to do with fundamentals of —

Horne: That's right. He was interested in the fundamentals of quantum mechanics. They were playing with one-particle interference with the neutron. And we started talking to each other over every dinner at that meeting, and we found out we very much liked each other's ideas about what's interesting. And I would talk a lot about two-particle stuff which he hadn't really thought about, and then from my previous half decade in two-particle physics, and he would tell me what they were doing with their one-particle interferometer which I was just starting to get into with the MIT group. So we just bumped into each other at this meeting in Sicily. And then when I came back home sometime in the next few months or before the year was out Cliff Shull at MIT said, holding up a letter, "Do you know someone named Anton Zeilinger?" And I said, "Oh yes, I met him at a meeting in Sicily." He says, "Well, he wants to come over for a postdoc. What do you think?" and I say, "Great. It'll be fun. You'll like him. And it would be fun to have him here. He is a very good person." And so that's how Shull and Zeilinger and I got started together.

Bromberg: So what year did he actually come for the postdoc?

Horne: So he came and spent — he brought his family, brought his wife Elizabeth, and he had two children at the time. I think possibly one; his oldest daughter was around. And they moved in and spent a year here in the Boston area. And so we started talking, you know, almost daily about what to do with this new toy. You know, what would be fun to do with this new neutron interferometer. We were not particularly talking about two-particle quantum mechanics at that time. Years later we started talking about it, but I can tell you the reasons. But that was years later. So I actually — I should summarize what we did during that decade. The idea was to find interesting things to do with this new interferometer, and that is basically following nature. Disturb it in some way, you know, rotate it, turn it on its side, subject it to various fields — things that would definitely cause a phase shift. Then we would do calculations to see what the phase shifts would be, and then we would do an experiment to see what they were. We didn't anticipate any surprises, but this was a whole new thing to do, subject neutrons to various perturbing influences and see if the quantum mechanics did what it was supposed to do.

Bromberg: In that whole group is there a lot of connection with what the quantum optics people are doing with their interference or are they just two separate worlds.

Horne: They were sort of separate. We didn't — by then quantum optics had gotten fairly sophisticated. I wasn't particularly knowledgeable about quantum optics. In fact I really knew very little about say what was going on in Rochester. Very interesting experiments were taking place with optics, and quantum optics was becoming a very rich field with a rich theoretical structure to it, and it was second quantized, it was quantum mechanics applied in a new area and I wasn't in that area. I wasn't working over there. What we were doing was sending these neutrons one at a time to this two-path gizmo and we — it was so basic we felt all we had to do was just take the de Broglie waves and do the calculation. The master equation for our work was the Schrödinger equation, but it was actually even more basic than that. We just needed to know the de Broglie wavelength and then work out it would change due to the perturbation we were playing with that year. So we rotated them and we tilted them and we perturbed them and there was always lots of interesting questions about the detail of, at this point, the de Broglie optics. Inside these crystals there were lots of interesting questions about how to work that out in detail. Because it wasn't just here's a place where the beam gets split; it was a thick region where it gets split, you know, inside this crystal. So there were lots of interesting things about the interferometer itself that needed to be worked out. And Danny [Daniel] Greenberger had been working on some of those questions once he started working with the Missouri group. Greenberger got involved with Overhauser, who was one of the people who first played with Sam Werner with the neutron interferometer in Missouri, and Danny saw some things from past experiments that puzzled him. Way back in the early sixties people had done electron interference. And one of the interesting things they had done with it was Aharonov-Bohm experiments. And Danny was, unbeknownst to me at the time, was looking back at those experiments and was puzzled about how they were able to see the effects they saw even though there were magnetic fields in the rooms. And he got really interested in these questions about electron interference. And that's how he got started thinking about one-particle interference, whatever the particle. Incidentally I had started reviewing some of those old electron experiments motivated by now doing the neutrons, and I came up on the same puzzle that Danny did, right? I thought, "Gee, I wonder how that experiment worked back in the sixties even though there were magnetic fields." And it turns out there was a compensating effect. And we both had independently worked on that, but we didn't know about each other. What happened was Danny was very much interested in what might be done with these neutron interferometers. He was specifically interested in the gravity experiments that had been done at Missouri. Somebody told him that a new person would be starting to do experiments with neutrons up at MIT, and so he came up to talk to these people, Cliff Shull. And that's where he ran into us. He came up and he was interested in basic quantum mechanics. Some of his interests were some experiments having to do with the nature of time and how it could be studied using a neutron interferometer. He had a background in relativity and basic — I don't know all these details. He still works on these sometimes even today.

Bromberg: Yeah, well of course he's on our list.

Horne: Yeah. So you should talk to him about the particular things that he was thinking about in going up to MIT to meet Shull. He wanted to know — It turned out Danny had written a paper back in the sixties that suggested that this gravity experiment be done, but he didn't know that it could be done. Right? He didn't know anybody to do it. He might have even come up to visit Cliff Shull early in the seventies, you know before these interferometers got created, with the idea that maybe he could make one. So this predates the Rauch group in Vienna and the Missouri group. He was interested in trying to make a neutron interferometer before anybody else starting making them. And then of course it turns out they showed up on the scene years later without his being involved, and so he started getting involved — first with the Missouri people and then with the MIT people. So that's the background that got me bumping into Danny. By '78 the three of us knew each other, Danny and Anton and me, and we were bumping into each other at MIT. Danny would come up from New York say for a weekend or for a week. Sometime in the summer he'd come for a couple of weeks and we'd talk about neutron interference and various questions that were coming up at that time about the instrument and about how to use it. So there are sort of two types of problems like getting a deeper understanding of what's going on in the crystals and thinking of new fun things to use it for. And so that's sort of the gist of the late seventies and all of the eighties.

Bromberg: And so you were not like Abner Shimony trying to work out any of the ontological or epistemological conclusions of Bell's theorem. That was not —?

Horne: That was over. That was over. But Abner would occasionally show up at MIT with his idea about what we might do.

Bromberg: I see.

Horne: That is, there were proposals for slightly modifying the Schrödinger equation. One of them came from Poland, introducing a nonlinear term, and Abner realized that this sort of proposal could be tested with a neutron interferometer. And so he came over and said, you know, we had various things we were doing and he said, "Here's one you might want to do," You can check this non-linear, non-orthodox term that people were thinking about sticking in the Schrödinger equation.

              And we did the experiment. Shull and his group did the experiment. That was one proposed by Abner. So you know I had ideas for things to do and worked with Shull's students while they went and did them, like a proposal on rotation. I spent a lot of time talking about rotating the thing, because the effect of rotation had been studied back in the 19th century with optical interferometers, and we thought, "Gee, we should do the neutron version."

Bromberg: I see. I didn't realize that.

Horne: And then there was also the question of putting matter inside the inter-
            ferometer to make phase shifts, but not just the matter itself, put the
            matter in motion. And there were famous experiments back in the 1850s
            — Fizeau in France, you know, had done the so-called Fizeau Effect
            where he studied the effect of moving matter in an optical interferom-
            eter in various experiments. Famous experiments from 1850. And so
            we said, one of the things I started thinking about doing was, "Let's do
            that with neutrons. Let's put the aluminum" — or whatever we were
            using as phase shifters — "and let's put it in motion and see the neutron
            Fizeau Effect." So that was one of the things I had worked on in the
            beginning. And you know Abner had his and various ones came up
            with various suggestions. So by '78 there were starting to be meet-
            ings about neutron interferometry and lots of experiments had been
            finished with Abner's suggestion and the Fizeau Effect and the gravity
            effect that Danny had been interested in and had actually got done in
            Missouri. So all these things were doable and were starting to be done,
            and we started meeting not only at Cliff's labs but at conferences that
            were taking place. And one of them that happened early, in the late
            seventies was at Grenoble (Bonse and Rauch 1979). And Danny went
            and Abner went and Shull went and I went. We all went there to talk
            about the things we were doing, each of us and together, and that's
            the first time I met Rauch, right? I knew about Rauch because he was
            Anton's mentor, he was his advisor, but I first went to Grenoble and
            ran into some of the other people besides Anton.

Bromberg: And that's in Vienna?

Horne: And they were in from Vienna, but it was in Grenoble that they were
            going to have this meeting, because Grenoble has a great source of
            neutrons, and years later that reactor would be used to do other experi-
            ments, many with Rauch involved in it, and Anton, where the neutrons
            were very slow and you had a much longer de Broglie wavelength. So
            it was at that meeting I presented my work on the Fizeau Effect and
            met some of the people from Europe who had a background in playing
            with neutrons. Some of them, their ideas on the thing with neutrons
            had come from Europeans who first built these instruments to do X-ray
            interference of the same wavelength. They were the ones who started
            the crystal interference business. But they were doing it with X-rays.

Bromberg: Was that an interesting cross-fertilization?

Horne: Well that's how the instruments came to be.

Bromberg: I mean from your point of view, did it enlarge the way you were thinking
            about things, or just didn't have all that much effect?

Horne: Well, that's how these instruments got started, from people who came
            from another direction. My basic interest from the beginning — I'm
            going to have to get me some sugar by the way because I have taken

my insulin for the day and have to start eating at regular times. Let me run and get some juice.

Bromberg: Okay, and maybe I'll get some water. [recorder turned off, then back on…] Let's see now.

Horne: So I think I've described — You've already talked to Abner, and so you know how we got together [on Bell experiments] and what we worked on and about Clauser being independently doing the same thing and then we got together. So since we started today I think I've described how the second part of my career, which was the one-particle neutron interference, got started and how the various players got together. Anton was already doing neutron interference as a student with Rauch, and I had joined the MIT group just looking for something fun to do, and then we both ended up working here off and on for fifteen years, Anton and I. He would come for visits and Danny would come up and join us.

Bromberg: And that got us up to about Grenoble.

Horne: Grenoble was just one of many meetings. This business went on for fifteen years. Nothing particular to report except you know they were just very fundamental experiments. That is, you know, we would do one and then move on and do another one.

Bromberg: Well of course the transition — I don't know if this is the next transition, but you get this transition to using entangled particles.

Horne: Okay, well that's the next chapter. I can tell you that story. So basically the first period is the Abner–Clauser two-particle entanglement specifically addressing Bell experiments and showed it up. Einstein's elements of reality don't exist. In the second period, the one which we were just describing —

Bromberg: Right.

Horne: — was one-particle interference and with various little neat, never-done-before experiments, but none of them worth singling out particularly. We could just do things that had never been doable before. But then there was a time when I went back to more than one particle and that started in '85. And it started because Anton was here on one of his many visits and we were sitting in Shull's lab, and one of the drawings that we always had in Shull's lab — it was almost like we had a tablet of them printed. It was a diamond. Right? And when we wanted to start talking about what we were doing with the current experiment that's been going on for years, we'd always point to the diamond because there is the two paths and we'd talk about what we were doing on each path. It was just the talking stage. So those sorts of figures were all around us, these single diamonds for a two-path interferometer. We got an announcement for a meeting to take place in Finland to (Lahti and Mittelstaedt 1985) celebrate the 50th anniversary of the Einstein-Podolsky-Rosen paper. And Anton knew I had been involved in the Bell EPR business but hadn't for about ten years been actively doing anything. By the way, during that ten years — just as an aside — during that ten years it had all of a sudden become widely known, that more people were interested in this Bell stuff then back when we started. I think there's a graph in this AJP collection of papers that shows how many people referred to the Bell paper, and there's like nobody all through most of the seventies and then by the mid-eighties it goes wow and it goes flying up, you know, like this.

Bromberg: Which of these AJP papers?

Horne: Yeah. There's a volume (Ballentine 1988). I think Ballentine is the Editor. It's at least fifteen years old now. It's a green paperback, 8 1/2 by 11 size. It's a collection of the key papers in this foundation business. It might be called — You haven't seen this?

Bromberg: I didn't know about this.

Horne: The whole foundations of quantum mechanics. Let me get you this. You should take a look at this book.

Bromberg: Absolutely.

Horne: Turn it off for a second.

Bromberg: Yeah. [recorder turned off, then back on...]

Horne: So there's '85, and we saw that this meeting was going to take place in Finland on the 50th anniversary of EPR, so Anton and I said, "Gee, it would be fun to go to that." But what can we present, right? Because we haven't been doing any two-particle stuff, what could we do that would be — And so we said, "I wonder if there is something about" — you know, I had been telling him about two-particle quantum mechanics, you know, of the simplest variety, where each of your two systems only possesses a two-dimensional state space, right? Two states, and what's an entangled state is this one goes with that one and then this one goes with that one that's entanglement. All the examples that we did with Clauser and everybody were so-called polarization where it's up-down, right? So Anton and I said, "I wonder if there is something we could do that would be a fusion of what we currently do." And we were looking at this diamond, and that old stuff, right? And we said, "Well, up-down in the old days, left to right, up goes with up, down goes with down, and we need two diamonds!" And this path is connected to that one and this one is connected to that one. So hey, so we started thinking about what came to be known to us as a two-particle interferometer. This source in the middle emits a pair of particles — and you're told that the two members of the pair go opposite. Right? That's the nature of this emitter.

Bromberg: Right.

Horne: It emits pairs that go opposite. But classically you say, "Oh that means either they go along those two legs of the diamonds or they go along those two legs of the diamonds as they retreat from each other."

Bromberg: Right.

Horne: But quantum mechanics says that's not what really happens. They go both this way and that way. So that's our entanglement. But it's entanglement of particles as they just travel through space, not something subtle, like the up-down or anything of polarization. So that's where the double diamond came to be, and we called it the two-particle interferometer. So we played with it for about an afternoon just on paper, and we saw that it was completely analogous — not surprisingly — to a Bell or Bohm or you know a standard polarization entanglement.

Bromberg: Now you had been deeply involved in that stuff at one point. Had he also?

Horne: No, he had not been involved in it, but we both sat there together and we just said, "Look. The two-particle quantum mechanics? We've been doing one particle for fifteen, for ten years. What's the connection? Can we make our connection between this one-particle quantum game we keep playing, and this?" or "Can we make a connection between that and this old stuff that I used to do." And it just, boom, it became that, right?

Bromberg: Yeah. And I should tell the tape that "this and that" means one diamond versus two diamonds.

Horne: Yeah. Two diamonds. So we worked it through and we found out that such an arrangement — we didn't know how to actually produce one at the lab, because we didn't know where to get a source that would emit pairs of particles in opposite directions.

Bromberg: Yeah, I noticed you were talking about positronium annihilation and —

Horne: Right.

Bromberg: You didn't think of cascades, atomic cascades?

Horne: We thought about cascades, but cascades don't have the desired oppositeness direction to them.

Bromberg: Oh. Okay.

Horne: The cascades are very fuzzy. That is, the particle goes this way and the other one can go anywhere over here with some distribution, right?

Bromberg: I see.

Horne: We wanted something that really was — But what we didn't know about was down conversion. We didn't know about that.

Bromberg: Yes. Which the optical community was at that point really beginning to.

Horne: And they were — completely unknown to us and vice versa. We didn't know about them; they didn't know about us. They were moving towards producing pairs of just the sort we were thinking about there on our piece of paper. So anyway, we gave the talk in Finland and people liked it. They thought it was interesting to see this —

Horne: That was '85. That year Danny had a meeting in the World Trade Center. Danny had a nice meeting. There must have been a couple

	hundred of us there, you know foundations of quantum (Greenberger 1986) mechanics. There's a nice publication for that, and it's also green.
Bromberg:	Yeah, and I've seen that.
Horne:	Right. By the way that's the same green paperback, same color as the other book I told you to look for. It's that same — this Ballentine is also that same dark green. We were all sitting there listening to the talks, and two people there talked about down conversion — [Yanhua] Shih and —
Bromberg:	Not the Mandel group?
Horne:	I think the Mandel group might have been there too, but I think there were at least two groups there that talked about down conversion, but we were just, you know, we didn't pay any attention. They actually were talking about sources that were just the kind we wanted, but we just glazed over and weren't paying enough attention. Anyway, so it wasn't until '87 that I happened to drop back by Shull's lab one day, some Tuesday, and there was a stack of Phys. Rev. Letters lying there that I hadn't looked at, months old most of them, and I was thumbing through one of them, and on one page I saw this picture that was a diamond inside a bigger diamond.
Bromberg:	Right.
Horne:	Right. And it was the picture from Ghosh and Mandel's paper. And I said, "Gee, that's our figure where you just unfold it," you know, have them opposite each other instead on top of each other, and make them the same size instead of one littler than the other? I said, "See, they're doing our experiment. There it is, right there. They're doing two-particle interference." So we had an idea independently of the whole world, but the whole world got without ever needing us, right? You know the [unintelligible word] two-particle interferometer. I wrote Mandel after seeing that paper and sent him my little paper from Finland, and he wrote back and says, "Your approach is very simple, very clear." I'd looked at — they had a paper where they talked about their theoretical underpinnings and what led them to do this experiment, and it was typically a 10- or 20-page Phys. Rev. Paper, heavy machinery —
Bromberg:	Lot of annihilation operators.
Horne:	Lots of annihilation. It was just like I couldn't even read it. I didn't even know what it was about, most of it? But our discussion was half a page. Well, we had exactly the same final result, we said it behaves just like they said it behaves, but our way of discussing it was just extraordinarily simple. And so he urged us to publish it. He said, "This is so much simpler than the way we describe it, you should publish it," and then he turned out to be the referee, you know.
Bromberg:	By the way, do you keep letters like that, like the one he —?
Horne:	It might have been on the phone. It might have been on the phone. Anyway, so that's the '89 paper with Abner [Shimony] and Anton

[Zeilinger] called "Two-Particle Interferometry," Phys. Rev. (Physical Review Letters 1989)

Bromberg: Did James Franson enter into this whole thing?

Horne: Yes, and about that time — sometime in there — we saw Franson's proposal, which was really quite remarkable because his was an ingenious new type of entanglement. Right? Whereas ours was just "this direction is entangled with that one and this one with that one," right? His was, if you have a pair — [phone interruption; recorder turned off, then back on...] Franson imagined a situation where the two particles of the down conversion source don't ever come together. There's just one branch going out one way and the other going out the other way, but each of those branches has two routes to get to the end of it — a long route and a direct route. So classically you could say, "Let's catch the particle simultaneously at the end of these branches. Let's catch them classically." They were produced at the same time. These down conversion particles are produced at the same time, so if you catch them at the same time you know it's a pair and it was produced at the same time. Classically you could say, "Well, if I catch them at the same time, that means both of them must have gone the long route through their respective branches — or the quick route through their Letters. which is actually a real experiment. All we're doing is just talking about it. And we pointed out that it's a complete analog to the old polarization versions and that therefore Bell's theorem applies. And subsequently Mandel and various other groups that had down conversion sources actually did some Bell experiments, you know. You know, it's sort of like the horse is already out of the barn. There's the Ghosh-Mandel, branches."

Bromberg: Right.

Horne: But quantum mechanically it's not either this or that; it's the old both and again, right? So there was a new type of interferometer, two-particle interferometer, but is completely different than the one we have described. Which were alternative routes. These are alternative routes where they never came back together, right? Now I liked that paper. That's a great paper, the Franson paper.

Bromberg: Did that interact at all with what you were doing or it was just a nice paper that was —?

Horne: No, it opened my eyes to that there must be lots of variations, that is there must be many ways to have these simple entanglements. Once we had broken out of the idea that a simple entanglement doesn't have to be a polarization entanglement. You know, that you can have simple entanglements that are spatial, and then Franson shows there's another way to have a simple spatial entanglement. It was a very exciting time, the late eighties, when I saw these various ways of making new kinds of entanglements. So anyway, you see our role of getting back into more than one particle, you know after, ten years of neutrons, was a conference paper in Finland that had no real effect. It had no real effect on anybody. Then the quantum optics people a few years later actually were doing the experiment, and then variations of it started popping up. And then so by 1990 there must be five or six groups around the world easily doing down conversion interference experiments.

Bromberg: And Shih was certainly doing a lot of interesting stuff.

Horne: And Shih was in there too, but he was in there from the beginning, right?

Bromberg: Well, from about the late eighties, because he just got his degree in the late eighties.

Horne: I think he was a —

Bromberg: He was a Carroll Alley student and I think he just —

Horne: I think it might have been Alley that was talking at Danny's meeting in '85. Remember I said there were several people who —?

Bromberg: Yeah.

Horne: In retrospect we now know we were talking about experiments that use down conversion, but we didn't notice. Yeah, I think that Alley — and Shih might have been involved with it — I think they did a Wheeler delayed choice experiment that they described at Danny's meeting, and I think the source was down conversion.

Bromberg: They certainly were working with down conversion from the start, those guys.

Horne: Yeah.

Bromberg: So somehow these two independent lines, neutron interferometry and optical, are really coming together in a way in which you're reading each other's stuff and getting ideas from each other it sounds like.

Horne: That's right, yes. That's right. In other words we were totally unaware of the trajectory that the quantum optics people were following in the middle of the eighties. We just weren't paying attention and weren't reading journals.

Bromberg: Even Abner Shimony wasn't paying attention?

Horne: Mm-mm [negative]. It was completely separate. So that brought us back. So now we have double interests. We are still pursuing our one-particle neutron interference experiments, although by this time they are starting to run their course. There are not many new ideas for "hey, let's do this with them." Anton continued after he — he continued and did some experiments with neutrons, single-particle neutron interference at Grenoble, and then since then in the last ten years, five years back in Vienna he's continued to pursue single-particle interference with atoms.

Bromberg: Mm-hm [affirmative]. Right, right.

Horne: He wasn't the first. I think Dave Pritchard at MIT and some of his —

Bromberg: Clauser did some atom interferometry.

Horne: Yeah. Later he did an atom too. There are several groups that have done atoms, but I remember it was in the late eighties that we first, when

Anton and I were still hanging around MIT, we visited Pritchard and his student then was named Martin and they had a beautifully functioning atom interferometer. So it was just the two-route game again just like with the neutrons but these were atoms. Anton has pursued that nonstop ever since. He went on to do a — he built his own atom interferometer and did a lot of experiments over the past five years. And then recently he did a bucky ball interferometer.

Bromberg: Oh.

Horne: Right. The particle that takes two routes is actually sixty or seventy carbon atoms, he did both types. It was together in a hollow sphere. It was a giant molecule.

Bromberg: I have a vague knowledge of it. I don't have a clear —

Horne: It's seventy carbon atoms all lumped together to make a hollow ball.

Bromberg: Okay.

Horne: That's the particle that now is going through the interferometer.

Bromberg: It's really extraordinary. You know I noticed that he's part of some European group in this — there's a recent volume that they put out on quantum information that I think Zeilinger is one of the editors of, and there is this European commission group to look at stuff like this. It sounds as if the European governments are interested in looking at this so that if it becomes a technology they'll have some leg up on it.

Horne: Yeah.

Bromberg: Does his belonging to a group like that have any influence on your collaboration or?

Horne: No. We still talk regularly. We haven't jointly published any papers for about three or four years.

Bromberg: What I'm really wondering is whether belonging to a group like that might push him towards applications and because you two are then linked it might be pushing you in any direction or another.

Horne: Well, since about 1990 when the more than two-particle entanglements came on the scene — which we didn't quite get to —

Bromberg: That's right. We're not up to that yet.

Horne: So maybe I should try to get to that. So we got back to two particles starting in '85, Anton and I did, and then we saw that the quantum optics people were already there, you know, at least two years later and actually doing the experiments. Sometime in about those years, like '88 or something, Danny asked one day — I think it was sitting right here [in Horne's kitchen] with Anton — he says, "Do you think there would be something interesting with three particles that are entangled? Would there be any difference, something new to learn with a three-particle entanglement?" You know, and so he was thinking of a polarization version, and of course immediately I started saying three diamonds, you know like from Mitsubishi. Because I always like to keep going back to space and stay away from polarization, right? So that's the first time — sometime around there is the first time I did the three diamonds.

Bromberg: Why do you like to keep away from polarization?

Horne: Well just because I, you know, as Feynman said, the ultimate mystery, the only mystery of quantum mechanics is the two-slit interference experiment.

Bromberg: Which is space?

Horne: And that's, you know, its equivalent. But if you talk to people, you know if you talk to general students or something like that where I teach, and you want to tell them any of these stories, polarization is sort of like "huh?' you know, "Spin? What?" They don't know what that is. But you know, did the particle go through this hole or did it go through that hole? So anyway, Danny said, "What if I considered more than two? What about three or four. Would there be something new there, something interesting?" And I think Anton and I encouraged him to pursue it, and he started working on it and he worked on it for like a year, part of the time here, part of the time on a sabbatical in Europe close to Anton. I wasn't intimately involved. Anton was probably closer involved but not actually working on it. He would talk to him occasionally. Danny would report back sometime over that year things like, "I have a Bell's theorem without inequalities" is the way he put it. "I can prove Bell's theorem without inequalities." Prior to that he was reporting things like, "Inequalities are popping out all over," right? And in between those two moments, between "inequalities are popping out all over" and "I have a Bell's theorem without inequalities at all", he began to suspect that the incompatibility that was working here between the Einstein point of view, tried out in a three-particle or four-particle context, the incompatibility is closer to the surface. You know, that it's not as hard to exhibit as it, say, was at Bell's original theorem.

Bromberg: So when you guys were talking around this table here were you already thinking that with three particles you could cast some light on Bell's theorem?

Horne: Oh, no, no.

Bromberg: That was just Greenberger.

Horne: No, he didn't have particularly in mind an improved or a different version of the Bell theorem. He just said, "I wonder if anything interesting happens with three-particle" — it could have been just purely a quantum mechanical question. You know, "What are the details of quantum mechanical entanglement for three particles?" with no particular regard to adding a new chapter to the Bell story. But what he actually found out was that he could prove the incompatibility of the Einstein point of view in the context of a three-particle entanglement. He could prove it quicker and shorter and sweeter than say the original Bell proof. And he said we should write it up. And I said well you know — He started writing and he'd get up to thirty pages and he'd say, "I have thirty pages and I still don't have it all in there." Because

	Anton and I kept saying, "Well we have lots of other little pieces that we should put together if we are going to write a paper." We just kept sitting on it. Nothing ever came out. We never wrote anything. And then one day Abner — Abner probably told you this story —
Bromberg:	No, he didn't.
Horne:	One day Abner went out, one day in '89 or sometime, I was with Abner at BU talking about something — down conversion probably, because by then we were very excited about down conversion — and he said, "Oh by the way, what's this thing you and Danny and Anton have proven?" And I said, "Well, what thing?" And he says, "Well," and he showed me this [N. David] Mermin article that was in Physics Today or something. And I said, "Oh. How did he know about that?" And it turned out that Danny had talked about the new three-particle, simpler way of proving Bell's theorem at George Mason University, and then he talked about it again at some talk in Europe, and people heard it. Right? Some people in England, I think that his name was Redhead.
Bromberg:	Yeah. Michael Redhead.
Horne:	Yeah, Michael Redhead heard it, and Mermin heard about it either directly or indirectly from somebody, and they looked at it and said, "Wow, this is a revolutionary breakthrough." So it's clear we needed to write something, right? Because everybody was commenting on it. People were commenting on something that had never been published, including in Physics Today. So you know like Mermin had a column or he has a column in the magazine. So we decided to get Abner's help, because I said well, if we're going to write this up, I decided, I convinced them that I had some pieces in mind, pieces back from 1969 that were never published. Specifically a step-by-step self-contained discussion of how EPR — you know, sentences that you can quote from Einstein, Podolsky and Rosen — lead you, you know, you can't avoid, getting to the form of theory that Bell was contemplating, you know hidden variable theory, and then shoving in the contradiction in the original Bell fashion. That had never been written or published anyplace, the detailed look at the EPR paper and you're led to Bell, right? You know, it's spelled out in detail, good for a sophomore. So we decided to put that in, and because it was going to involve so much Bell physics we decided — and since Abner was the world's resident authority on Bell physics — we decided to get Abner to join in our write-up. So instead of just Danny and Anton and I writing up this unpublished stuff the (Bell's Theorem Without Inequalities 1990) four of us wrote it up. And that's the 1990 American Journal of Physics paper.
Bromberg:	Right.
Horne:	Called "Bell's Theorem Without Inequalities." And that has several parts to it. One is a review of how EPR leads to these hidden variable theories, a review of Bell's way of showing that it can't be done, then

Danny's new proof in the context of four [particles]. He originally did it with four. Then we pointed out that it could equally well be done with three, and then when we switched to three we switched from polarization to space so we could exhibit the Mitsubishi triple diamond, right? Gives us a chance to put that out there. And then Abner put in a section about, is a new experiment called for. What we really felt was no, you know. I mean, the experiment had already been done that shows that you can't have these elements of reality and you don't need an extra particle to show you can't have them. That was a point of some controversy. A lot of people outside of our group kept thinking, "Gee, this really means we can do a better experiment," but I've never been convinced of that. Because the proof of incompatibility is so brief you'd think that would mean you can do a better experiment, but I don't think that follows. Right?

Bromberg: I see.

Horne: Oh, I should mention there's a whole world out there of what we call loophole busting. That is, there are all these people who just say, "You know, I can still keep the Einstein point of view in the face of quantum mechanics because your detectors don't count all the particles, and I can come up with this pathological scheme."

Bromberg: Yeah.

Horne: So one of the things that people had in mind was maybe the three particles would be a better way of blocking some of the loopholes. And people are still contemplating that. In fact Danny and I currently are working on something along those lines.

Bromberg: Are those people who don't feel that there's been closure yet on the Bell theorem, is that kind of a lunatic fringe or is that quite a serious group of physicists or how do you —?

Horne: My personal feeling is it's closer to the first. In the quantum mechanics, the quantum mechanical predictions for these multi-particle entanglements are confirmed to like ten standard deviations beyond where the cutoff point is. But you know, you can come up with these pathological things where all kinds of screwball things happens because you don't count all the particles, and if you strain hard enough you can say, "Look. You haven't shot down the Einstein point of view yet."

Bromberg: Yeah, okay.

Horne: It's not a field I'm particularly interested in.

Bromberg: Now I'm a little confused because — By the way, have you eaten what you need?

Horne: No, I think the juice ought to help me, and I can give us a sandwich in a second. So what was your —? Something confusing?

Bromberg: When is the GHZ paper, the one that everyone refers to? That's not the one you were just talking about?

Horne: That's the first published version, and it's actually GHZ and Shimony.

Bromberg: Oh. Okay.

Horne: Danny wrote and put Anton and my name on a paper for that George
            Mason proceedings. There's a meeting at George Mason and he put
            out a little paper. That one had circulated a little bit, but it's not very
            clear.

Bromberg: No, it's not. That's the one in the volume edited by Kafatos I
            think (Greenberger et al. 1989).

Horne: Yes. Right.

Bromberg: Yes. And it wasn't very clear and — I mean at least when I read it.

Horne: Right, yeah, that paper doesn't do a lot for me.

Bromberg: And I wondered whether you got much reaction to that paper because
            I had trouble with it.

Horne: Well it didn't have any circulation, you know, because it was just a
            conference. I don't know whether it was one that Mermin saw or
            Redhead saw it. I think Mermin said he had just heard Danny talk
            once. So, as far as "the GHZ paper," there is none, except that George
            Mason conference paper.

Bromberg: Okay.

Horne: We did later, say in '93, we wrote, we tried to describe, several of
            these developments, you know, the three-particle GHZ way of proving
            Bell's point, and also talking about some of these really remarkable
            two-particle experiments that were made possible by down conversion.
            So we wrote a Physics Today article, a news article. That's in '93. It
            was called "Multi-Particle Interference," right, and it had a few — it
            mentioned some Bell stuff in passing, but it wasn't just Bell physics.
            See, you've got to remember, I thought Bell physics was done in 1974
            or '73. I wasn't ever particularly interested in pursuing Bell physics
            anymore. When Danny kept saying after that year, playing with three,
            "Hey, Bell's inequalities are popping out all over," or "Hey, I can prove
            the incompatibility without inequality at all," to tell the truth I sort of
            glazed over. You know, it's a lot of, "Oh, that's good. You can work
            that out."

Bromberg: I see. [laughs]

Horne: "In other words, the first person who really said, "Wow, this really is"
            — You know, Danny couldn't clearly present it as something revolu-
            tionary. Danny wrote that George Mason thing. The first person who
            actually trumpeted it was Mermin. Mermin said, "Wow, look at this!"
            He paid more attention to it than I did. So that takes us from…

Bromberg: We're sort of up to the Physics Today paper of '93 right now
            (Greenberger and Michael 1993).

Horne: Now so the next thing that came up that I was involved in with Anton
            was, "Can we actually make in the lab a three-particle entanglement?"
            Right? That's what we talked about a lot all through the early and mid-
            nineties, "Can we actually make one? Is it possible to produce one
            on a tabletop?" And so I think that's the last paper that I wrote that
            I co-authored with Anton. I think it's '98. It's called "Three-Particle

Entanglement"? The content — I don't know the title — is how to produce a three-particle entanglement in the laboratory. A proposal for making three-particle entanglements. I don't know exactly the title. And then a year or two later they actually carried it out.

Bromberg: In his lab.

Horne: In his lab. Right. But I wasn't part of that, because I wasn't over there involved in the experiment or anything, but I was involved in the proposal as to how to do it. And then that's basically it.

Bromberg: And you're also interested in the physical origins of entanglement. Isn't that something that you worked on?

Horne: Yeah. Well, yeah.

Bromberg: At what point is the entanglement occurring?

Horne: Yeah, well, that's one way to put it. The question that we've often asked is, "Who ordered this?" — as Rabi used to say about some experiments fifty years ago with molecular beams and they'd get a surprising result and he would say to his collaborators, "Who ordered this?" when they saw this surprising result. I've always liked that expression. "Why did things behave this way?"

Bromberg: Right.

Horne: And that's not a very — Not many people think that that's a worthwhile question.

Bromberg: Really?

Horne: In other words, I think most people say, "Well, we know the quantum mechanical rules. The rules contain this aspect. Namely that entanglements can exist", right? "And we know the quantum mechanical rule. It's called superposition principle." Right? And that's what leads to these entanglements. You realize that entanglement is nothing but the superposition principle applied to a more-than-one-particle system.

Bromberg: Right, right.

Horne: That's all it is.

Bromberg: This combination of states or that combination of states.

Horne: So somebody says, "Well so why do we have this?" and they'd say well, "Superposition principle. You add these things together." And you know, Feynman says in his lectures that that's the only mystery.

Bromberg: Yeah, but Feynman was content to live with that mystery.

Horne: That's right. He goes on to say, "You can't answer it." He says the only mystery is the two-slit experiment, and that just means the only mystery is the superposition principle. That's the only thing about — that's the heart of quantum mechanics is that "Thou shalt add these amplitudes and you don't add probabilities," right? You add these amplitudes.

Bromberg: But it seems to me that —

Horne: And he said, "Don't ask what's behind it. You won't find an answer as to why it's like that. It's just that's the way it is." So when someone asks the question, when someone considers the question I just described, most people would say they're not sympathetic to that question. And by the way I have no answers. But it just sort of struck me as something — and it strikes Anton the same way — that there must be a reason. There must be some deep reason why these sort of strange linkages are a necessary part of the world. I mean, why do they link up like this? That glosses over the, that theory glosses over that. The answer is superposition principle. So what we're asking is, "Well then why the superposition principle?" And most physicists don't quantum mechanics would say, "Well, you're not going to answer that question." It's just that's the rules. You see what you're dreaming of is deriving quantum mechanics. Right?

Bromberg: Yeah. Is it deriving from quantum mechanics or understanding at what point —? I don't know. Does it —?

Horne: At what point does it have to show itself. Yeah.

Bromberg: At what point does the entanglement —

Horne: Have to show itself, right? Yes.

Bromberg: And what's the physical thing that's correlating the two? That seems to be a question physicists would be willing to entertain.

Horne: Yeah.

Bromberg: That's what I thought you were doing in that experimental meta-physics volume that was I guess one of the volumes dedicated to Shimony [Horne, 1997].

Horne: Right. What was my paper in there?

Bromberg: I thought you were trying to understand what was the origins of entanglement inside a crystal. I don't have any notes on it here, but as I remember it you had various sources within this crystal because it had finite dimensions that were–

Horne: Oh yes. Oh, oh, I know that paper. Nobody knows much about it. Frankly nobody knows that paper. Yeah. I imagine that every spot in the source, every spot is a source of radiation, sort of like Huygens, right, you know? Every spot on the previous wave front is the source of another little wavelet. Except in that paper I was imagining that every spot in the source material, in the source medium, is a source of pairs of spherical waves. Right? Two-concentric wavelets coming out.

Bromberg: Right.

Horne: Oh, yes. Oh, I think that's a very nice paper. If you then sum over the whole source, you know to see what the total output will be from all of those little sources, you derive momentum conservation. In other words, if the incident radiation that's stimulating the medium comes in this way, then when the two particles are produced out of all of these wavelets adding together, if this one goes off at 15 degrees that way, this one will be going off at 15 degrees this way. Right? If this one goes off at 20 degrees, this one will come off at 20 degrees. So I was struck that by using a sort of a two-particle Huygens construction. I could derive conservation of the momentum, right? I was deriving it, right? And that's classical physics, but I was deriving it from a bunch of little wavelets.

Bromberg: In fact the whole thing that really was striking me about that paper and one other is that there is no wave particle duality somehow in what you're doing. The waves and the particles are so meshed in conception that you are sort of moving back and forth from wave to particle without —

Horne: Exactly. I never appreciated this duality. It's just like quantum mechanics says there is nothing but these waves as far as your basic machinery and what you've got to calculate, and then you just use those amplitudes at the end to say, "Well, will I catch a particle here?" or "Will I catch one there?" or "What's the probability of catching one here or there?" So I get your point. There's nothing there except these little amplitude wavelets, right, and then just add them up together. I sort of liked that paper, but I've never done anything with it. I doubt if anybody has ever seen it or it did anything for anybody.

Bromberg: [Asks about Yenhua Shih's bi-photon concept] — I never understood where that fits in the way in which people think about these things, whether this is just a little bee in his bonnet that is not generally thought about or whether it's a really important idea that everybody picks up on.

Horne: Well, it struck me as a valid way of describing entanglements. It's just a verbal way of capturing what we were already reviewing a while ago, that an entanglement is — system A has this state times system B has this other state, plus another product like that, right? Or it's, like I said, it's just superposition. Now when a pair of particles have that kind of a state, neither one separately has a state. Right?

Bromberg: Yeah, right. Absolutely.

Horne: In other words it's just there and you say, "What's the state of those two things over there?" and you'd say, "It's this superposition," right, "It's this superposition." Well what's the state of just the object that's the one over on the left? It doesn't have a state. Right? So I think when Shih says you can't speak of these as two objects, it's just one object, it's just another way of saying that, right? In other words you know the — I was always very sympathetic to his wanting to emphasize that repeatedly. It seemed to be a valid way to present it verbally. It's sort of like his version of — you've already run into my favorite, is "both and" as opposed to "either or." Right? You know, that is, whether you are doing one particle or two particles in quantum mechanics what the superposition principle says — and the superposition principle is all the entanglement is, right? — what it says is, unlike classical where you have either this happen or that happen, what quantum mechanics says is that they actually both happen. Right? Simultaneously. That seems to be what she is saying. In other words I didn't see it as a variable menace and groundbreaking insight into learning new things about entanglement; it's just his verbal way of stressing how strange they are.

Bromberg: Okay.

Horne: It's his way of describing them, right? It's a single thing, you would say.

Bromberg: I'm going to ask you one final question. The quantum optics people were using this whole mechanism. We already talked about it; that derives from Roy Glauber, with annihilation operators and such. As your two fields began to interact, did the neutron people also begin to use all these?

Horne: No. No, because —

Bromberg: Because neutrons don't get annihilated and —?

Horne: The quantum mechanics we did with neutrons was the most elementary variety of wave mechanics. You know, that is, the propagation of the single neutron through the contraption you had set up for it in the lab was just described by a single scalar wave, you know, an amplitude really — a quantum mechanical amplitude — and it's physical interpretation is the square of it in any spot is proportional to the probability of finding the particle there. Just a standard Born interpretation of a quantum wave. So that was the extent of our quantum mechanics, right? It was just, that's all we needed were these waves. There was no sophisticated annihilation machinery, right? There were no operators like in quantum optics. None of this heavy machinery was needed for all the things we did. We could get a complete account of the experiments with just this simple, elementary way of working. So then when we started talking about pairs, both originally on our own but then later saw the quantum optics people were doing pairs, we personally from our background didn't see any need for the heavy-duty calculus. I have a notebook from the late eighties where I went through like a half a dozen of the [Yanhua] Shih and Mandel experiments, and I would use my simple little way of calculating the final formulas. And I'd get all the same formulas, right? So I could show just by example after example. I didn't need this machinery to get to the same answers. So I not only don't use it, I don't need it for the things I try to do. Now I don't know how far that might go. I might hit a wall sometime, right? I might hit a wall and say, "Gee, I can't go any farther unless I adopt the heavier machinery." But I never ran into it as far as I got that I needed it. And also I'm not an expert on it. You know, so it's fortunate that I didn't need it because I didn't know much about it.

Bromberg: Well, if you needed it you would become expert on it so you don't have to worry about that. Well good, that's the end of my questions for today.

Horne: So you had asked earlier how was it doing physics at Stonehill, and I think it's been perfect for me, given my temperament. After I finished grad school and had the opportunity to have that really great thesis topic, because, of Abner, then I needed a job. And Bob Cohen, as chairman of the physics department at the time, got lots of letters from people in the neighborhood in Eastern Massachusetts. You know, in small colleges when they had an opening they would often just send a letter to department chairmen in the area, you know, at big universities like BU or Northeastern or Brandeis. So Bob Cohen got such a letter from the then only physicist at Stonehill, which was Chet Raymo, and said, "Gee, this looks like it might be a good place for you to look for a job." So I went down and interviewed and got that job. Do you know Chet Raymo?

Bromberg: No. I know the name Raymo, but not Chet.

Horne: He was the physics teacher there in the sixties, and he wanted to do other things and so he talked the administration into hiring me and I would do the physics and he could continue to do things that he wanted to do more, which was write, write popular books on nature and teach students about writing and he ended up writing a column in the science section of the Boston Globe. It's been going out for twenty-five years.

Bromberg: Oh.

Horne: So Chet Raymo was the physicist at Stonehill and hired me so he could do the things he was more interested in.

Bromberg: But I mean you don't have to teach like four courses a term?

Horne: I do.

Bromberg: And it's not exhausting?

Horne: I did it ever since 1970, so I've now been doing it thirty-two years. I was teaching general physics, one for math and chemistry and one for biology, and I was free in my third course to do whatever I wanted. Often I've taught courses for general interest, you know general students, about quantum mechanics or about relativity or about both. Or occasionally I would do an intermediate course for somebody who wanted to do some physics beyond general physics and do something at the intermediate level. Okay, there was no major. And I very diligently never started one, right? So I thought it was good for me to not have the politics of a high-pressure university physics department working over me, you know, where you know you have to produce or you don't get tenure. And for me it was best to — I had my friends, Anton and Danny and Cliff Shull and Abner, and they were all connected to major, you know, bigger places, but they were my contacts to the real world of physics and it's been perfect for me. I've been involved in some, what I thought were exciting developments in foundations of quantum mechanics. But I was never under any pressure, you know. I taught at this school where I didn't have any majors. I was just a service department teaching general physics for the chemistry students and the biology students, but I didn't have any students.

Bromberg: How big is that school? How come they have no majors?

Horne: Well, it was a couple of thousand students, and to have a real physics
              department I think you need at least four or five professors to start an
              undergraduate physics major program, and I just didn't think — you
              know, there were enough physics major programs in the country. If
              occasionally a student came and said, "Gee, I just really like physics
              after taking your course," I would say, "Well, you should go somewhere
              else." I'd just tell them, "You should transfer, switch to another school."
              So it's been perfect for me — a low-pressure place where I was the
              physics department. Right? So it couldn't have been better.
Bromberg: That's interesting.

References

Ballentine LE (ed) (1988) Foundations of quantum mechanics since the Bell inequalities. American
    Association of Physics Teachers, College Park MD
Bonse U, Rauch H (eds) (1979) Neutron interferometry: proceedings of an international workshop
    held 5-7 June 1978 at the Institut Max von Laue–Paul Langevin, Grenoble, France. Clarendon
    Press, Oxford
Greenberger DM (ed) (1986) New techniques and ideas in quantum measurement theory. Ann New
    York Acad Sci 480
Greenberger DM (1990) Bell's Theorem Without Inequalities. Am J Phys 58(12):1131–1143
Greenberger DM, Horne MA, Zeilinger A (1989) Going beyond Bell's theorem. In: Kafatos M (ed)
    Bell's theorem, quantum theory and conceptions of the universe. Kluwer Academic Publishers,
    Dordrecht, pp 69–72
Greenberger DM, Horne MA, Zeilinger A (1993) Multiparticle interferometry and the superposition
    principle. Phys Today 22–29
Horne MA (1997) Two-particle diffraction. In: Cohen RS et al (eds) Experimental metaphysics.
    Kluwer Academic Publishers, Dordrecht, vol 193 of Boston Studies in the Philosopher of Science,
    pp 109–119
Horne MA, Shimony A (1975) Proposed neutron interferometer observation of the sign change of
    a spin due to 2 precession. Reprinted as Ch. 6 in Abner Shimony, Search for a naturalistic world
    view, vol II. Cambridge University Press, Cambridge, UK (1993)
Horne MA, Shimony A, Zeilinger A (1989) Two-particle interferometry. Phys Rev Lett
    62(19):2209–2212
Lahti P, Mittelstaedt P (eds) (1985) Proceedings of the symposium on the foundations of modern
    physics. World Scientific Publishers, Singapore

Chapter 3
Laboratory-Space and Configuration-Space Formulations of Quantum Mechanics, Versus Bell–Clauser–Horne–Shimony Local Realism, Versus Born's Ambiguity

John F. Clauser

Introduction—What Quantum Mechanics is Not

A commonly asked question is "What is quantum mechanics?" Perhaps an easier question to answer is "What is Quantum Mechanics Not?" One answer to the first question that is commonly given is that "Quantum mechanics is a description of the world of the very small". Given observation of quantized magnetic flux, quantized vortices in liquid helium, and the Aharonov and Bohm effect, all over square-centimeter-sized areas, and especially given observation of two-particle quantum mechanical entanglement that has been observed to extend over distances of 133 and >1500 km, that answer also seems rather difficult to buy. Another answer that is commonly given is that "Quantum mechanics describes nature non-deterministically". Local Realism was first formulated by Clauser and Horne (1974), as an alternative theoretical framework to that of quantum mechanics. Like quantum mechanics, Local Realism includes theories that allow non-deterministic evolution. Quantum mechanics is thus hardly distinguished from Local Realism by that definition. Yet another answer is that quantum mechanics provides a description of nature in terms of "waves of probability (causally) propagating in space–time." That answer also will be shown here also to be incorrect.

A survey of more than thirty quantum mechanics and quantum field theory textbooks and review articles reveals two distinctly different schools of thought of what quantum mechanics is. These books are found to promote three very different formulations of quantum mechanics. One is quantum mechanics formulated in laboratory space, the second is quantum mechanics formulated in configuration space, and the third is quantum field theory, which is formulated in a mix of these of these spaces,

J. F. Clauser (✉)
817 Hawthorne Drive, Walnut Creek, CA 94596, USA
e-mail: john@jfcbat.com

© Springer Nature Switzerland AG 2021
G. Jaeger et al. (eds.), *Quantum Arrangements*, Fundamental Theories of Physics 203,
https://doi.org/10.1007/978-3-030-77367-0_3

but differently by different authors. It is the purpose of this paper to describe, characterize, and contrast the configuration space and lab space formulations of quantum mechanics and quantum field theory with each other and with Local Realism.

The first of these schools is the laboratory-space school of quantum mechanics, or lab-space school, for short. In that school, Schrödinger's equation is formulated directly in lab space. Lab space is the ordinary three-dimensional space in which we live. The lab-space school's complex valued wave function $\Psi_{lab}(\mathbf{r}_{lab}, t)$ propagates causally as a wave in lab space. The real-valued probability density, $|\Psi_{lab}(\mathbf{r}_{lab}, t)|^2$, also propagates in lab space. It is described as propagating similarly to a classical field, except that it is subject to a somewhat mysterious "statistical interpretation", which was invented by Max Born. Via the lab-space school of thought, one is led to envision "waves of probability". That "interpretation" portrays $|\Psi_{lab}(\mathbf{r}_{lab},t)|^2$ as a spatially dependent "probability density for finding the particle" at the point, $\mathbf{r}_{lab}$, that flows like a wavy fluid in space–time. Schiff (1955, p. 18) describes the waves and fluid as being similar to sound waves. Messiah (1961, p. 223) notes that "In the classical approximation, Ψ describes a fluid of non-interacting particles …, the density and current density of this fluid are at all times respectively equal to the probability density P and the probability current density $\mathbf{J}$ of the quantum particle at that point." Schiff (1955, p. 344) discusses dividing up all of (lab) space into cells and specifying the value of Ψ_{lab} at each cell. Schiff (1955, p. 348) says "*This application implies that we are treating* Eq. (6.16) [Schrödinger's equation in lab space] *as though it were a classical equation that describes the motion of some kind of material fluid.*"

Born asserts that this probability density may also be interpreted as the particle number-density of a beam of particles traveling through lab space. He famously showed that it may be used to demonstrate conservation of particle number in Rutherford scattering of the beam. Under the lab-space formulation the wave function's value (whatever it means) and the associated probability density's value are correspondingly defined at every space–time point, $(\mathbf{r}_{lab}, t)$, within the laboratory.

The quantum mechanics textbooks by Born (1933, 1935, 1969), Schiff (1955), Dicke and Wittke (1960), Merzbacher (1961, 1970), Eisberg (1961), Eisle (1964), Feynman and Hibbs (1965), Feynman's (1948) seminal review article, and French and Taylor (1978) all formulate quantum mechanics in lab space. Curiously, none of these books explicitly specifies or even calls notice to this important aspect of its description.[1] The wave-function's argument-space choice can be taken from context and from the supporting illustrations.[2] Typical wave and particle motions are depicted

---

[1] By exception Feynman and Hibbs (1965) say the book is based on a review article by Feynman (1948), titled "*A space–time approach to non-relativistic quantum mechanics.*" Schiff's (1955, p. 348) second quantization of $\Psi_{lab}(\mathbf{r}_{lab},t)$ admits that it treats the single-particle Schrödinger's equation "*as though it were a classical equation that describes the motion of some kind of material fluid.*".

[2] Authors that formulate quantum mechanics in configuration space almost always call the reader's attention to the argument space that is being used. On the other hand, authors that do not specify configuration space, instead generally use a lab space formulation. Merzbacher (1961, 1970) only introduces configuration space when he discusses quantum field theory (in his second edition), but

in Figs. 3.1, 3.2, 3.3, 3.4, 3.5, 3.6 and 3.7 that are reproduced from these books. French and Taylor (1978) further describe a large number of experiments whose results seem

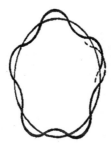

Fig. 18.—Motion of a wave in a circle; a single definite wave form is only possible when the circumference of the circle is a whole number of times the wave-length.

Fig. 3.1 Lab-space matter waves propagating around a nucleus, reproduced from Born (1933, 1935, 1969, Fig. 18, p. 132)

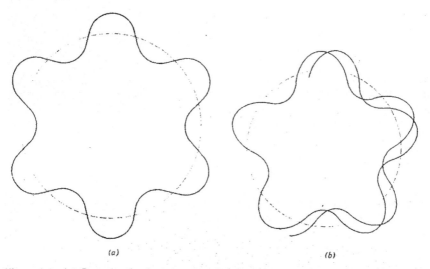

(a)   (b)

Figure 1.1. (*a*) Constructive interference of de Broglie waves in an atom distinguishes the allowed stable Bohr orbits. (*b*) Destructive interference of de Broglie waves in an atom disallows any orbit which fails to satisfy the quantum conditions.

Fig. 3.2 Lab-space matter waves propagating around a nucleus, reproduced from Merzbacher (1961, 1970, Fig. 1.1, p. 6). Eisberg (1961, 1967, Figs. 6–7, p. 152) also shows a very similar figure

---

not until then (see also his quote in the section "Born's Argument-Space Ambiguity"). His vague lack of specificity is traced to Born's argument-space ambiguity, discussed in the section "Born's Argument-Space Ambiguity".

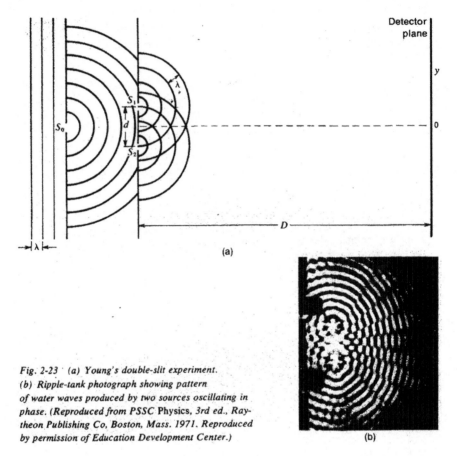

(a)

Fig. 2-23 (a) Young's double-slit experiment.
(b) Ripple-tank photograph showing pattern
of water waves produced by two sources oscillating in
phase. (Reproduced from PSSC Physics, 3rd ed., Ray-
theon Publishing Co, Boston, Mass. 1971. Reproduced
by permission of Education Development Center.)

(b)

Fig. 3.3 Young's two-slit *Gedankenexperiment*, and photograph of wave motion in a ripple tank, reproduced from French and Taylor (1978, Fig. 2.3, p. 90)

to demand a lab-space explanation for the evident wave-like properties. By using lab space, the motion of fields and particles is readily visualized and understood. Abstract mathematics with no clear conceptual connection to the physics is generally avoided. Indeed, it was the stated intention by French and Taylor (see their preface) to use a description that presents a "clean story line". Schiff (1955) says, "we shall try to make the theoretical development seem plausible rather than unique."

Figures 3.1 and 3.2, are reproduced from Born (1933, 1935, 1969, Fig. 18, p. 132) and from Merzbacher (1961, 1970, Fig. 1.1, p. 6), respectively. They depict the motion of waves propagating in lab space around the nucleus of a hydrogen atom. Eisberg (1961, Fig. 6–7, p. 152) also offers a very similar figure, following his

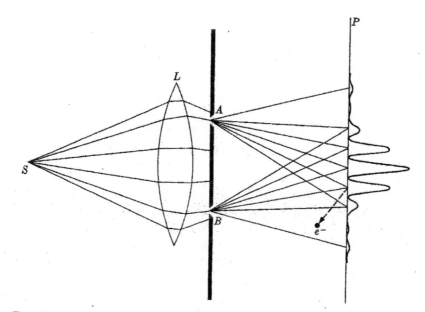

FIG. 2–1. Schematic representation of Young's interference experiment, illustrating the wave-particle duality paradox.

Fig. 3.4 Particles and waves propagating in lab space in Young's two-slit *Gedankenexperiment*, reproduced from Dicke and Wittke (1960, Fig. 2.1, p. 21)

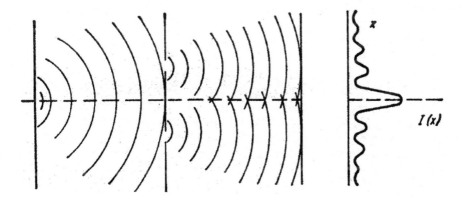

Fig. 3.5 Young's two-slit *Gedankenexperiment*, reproduced from Feynman and Hibbs (1965, Figs. 1-3, p. 5) showing propagation of Ψ waves in lab space

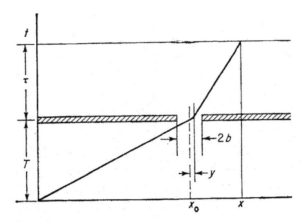

Fig. 3.6 Lab-space geometry used for calculating path integral solutions to the Helmholtz equation, reproduced from Feynman and Hibbs (1965, Figs. 3–3, p. 48)

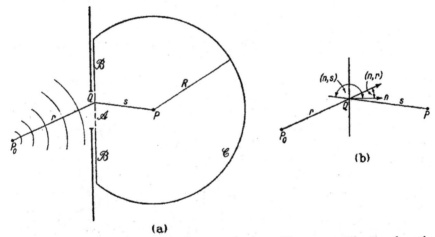

(a)

Fig. 8.3. Illustrating the derivation of the FRESNEL–KIRCHHOFF diffraction formula.

(b)

Fig. 3.7 Lab-space geometry used for calculating path integral summation solutions to the Helmholtz equation via the Fresnel-Kirchoff diffraction formula from classical physical-optics theory. It is reproduced from Born and Wolf (1959, 1987, Fig. 8.3), *Principles of optics*. It is for comparison with Fig. 3.6

promotion of deBroglie's pilot-waves lab-space formulation of quantum mechanics[3] (pp. 141–146). Figure 3.3 is reproduced from French and Taylor (1978, Fig. 2.3,

---

[3] deBroglie's pilot-waves formulation of quantum mechanics provided strong motivation for the development of Bell's Theorem and Local Realism. Unfortunately, it is soundly refuted by experimental evidence against Local Realism. See the section "Bell–Clauser–Horne–Shimony Local Realism".

p. 90). It depicts propagation in lab space of matter-waves through Young's two-slit *Gedankenexperiment*. It also compares this to a photograph of wave motion in a laboratory ripple tank. Figure 3.4, taken from Dicke and Wittke (1960, Fig. 2.1, p. 21), depicts motion of particles and waves propagating in lab space in Young's two-slit *Gedankenexperiment*. Figure 3.5, reproduced from Feynman and Hibbs (1965, Fig. 1–3, p. 5) depicts wave motion in lab space in Young's two-slit *Gedankenexperiment*. Figure 3.6, also taken from Feynman and Hibbs (1965, Fig. 3–3, p. 48), shows the lab-space geometry used by them for calculating path integral solutions to the time-independent Schrödinger's equation, which is the same partial differential equation as the Helmholtz equation. Figure 3.6 may then be directly compared with Fig. 3.7, which is reproduced from Born and Wolf's *Principles of Optics* (1959, 1987, Fig. 8.3). It shows the same lab-space geometry that they use for calculating path-integral solutions to the Helmholtz equation via the Fresnel-Kirchoff diffraction formula from classical physical-optics theory (in lab space and with no particles evident).

The lab-space formulation is reviewed below in the section "The Lab-Space Formulation of Quantum Mechanics". The formulation in lab space of the single particle Schrödinger's equation is described in the section "Single particle Schrödinger's equation in lab space". The section "Born's Probability Density and Conserved Probability Current Defined in Lab Space" describes the calculation of probabilities in lab space and Born's conserved probability current that also flows in lab space similarly to the flow of a fluid. This current was invented by Born to explain particle flux conservation in Rutherford scattering. It is only capable of being formulated in lab space, and is a concept that only makes sense in lab space.

The second school is the configuration-space school. It is described below in the section "The Configuration-Space Formulation of Quantum Mechanics". Schrödinger's equation is formulated using a highly abstract mathematical space. It describes a very general quantum mechanical system. That space is called configuration space. Its complex valued wave function $\Psi_{config}(q_{1,config}, ..., q_{k,config}, s_{1,config}, ..., s_{k,config}; t)$ is very different from $\Psi_{lab}(\mathbf{r}_{lab}, t)$. Rather than specifying the position within the lab where the wave function is to be evaluated, the wave-function's arguments instead specify the various degrees of freedom of the described system. For a system composed of N particles, the wave function, $\Psi_{config,N}(\mathbf{r}_{1,config}, \mathbf{r}_{2,config}, ..., \mathbf{r}_{N,config}, t)$, has arguments that specify then N positions, $\mathbf{r}_{1,config} - \mathbf{r}_{N,config}$, of these N particles. The arguments may be either continuously varying or discretely varying, and may include non-classical degrees of freedom, like spin, $s_{1,config}, ..., s_{k,config}$, and isotopic spin.[4] A discretely varying argument with M allowed values, in turn, may be used as an index, so that Ψ_{config} may be considered to be a vector-valued function with M components. Quantum mechanics textbooks that promote this school include the books by von Neumann (1932,1955), Landau and Lifshitz (1958, 1965), Messiah (1961) and Bjorken and Drell (1964). Interestingly, configuration-space-school textbook authors rarely acknowledge the existence of the lab-space formulation of quantum mechanics, and vice versa. Noteworthy exceptions are Dicke and

---

[4] See Bjorken and Drell (1964, p.222).

Wittke (1960) and Merzbacher (1961). Their comparisons of these formulations are described in the section "Born's Argument-Space Ambiguity".

There are also many configuration-space based quantum mechanics textbooks, which may be classified as applied quantum mechanics textbooks.[5] These books primarily discuss N-particle systems, whereupon this later feature forces them to use configuration space. Such books include the works by Pauling and Wilson (1935), Bethe and Salpeter (1957), and Condon and Shortley (1964). This latter group does not attempt to "explain" the N-particle Schrödinger's equation. Instead, the formalism is simply accepted as given and useful in calculating results that match experiment.

The configuration-space school is described in the section "The Configuration-Space Formulation of Quantum Mechanics". There are many additional configuration-space schools of thought to be found in various quantum mechanics books than are treated here in section "The Configuration-Space Formulation of Quantum Mechanics". Von Neumann's Chap. 1 formulation, listed therein as *The Original Formulations*, is followed in the section "The Configuration-Space Formulation of Quantum Mechanics", along with that by Messiah (1961). Other configuration-space formulations that are not discussed, for example, include Heisenberg's matrix mechanics, outlined very nicely, for example, by Condon and Shortley (1964), Dirac's (1930, 1935, 1947) formulation of quantum mechanics in Hilbert space, and others. Each formulation is proclaimed by its promoters to be mathematically consistent with all of the others, and each is a progressively more mathematically abstract. Correspondingly, each is progressively more difficult to visualize than the others. A discussion of all of these other related formulations goes beyond the scope of the present article.

Configuration space is defined in the section "Configuration Space". The section "Schrödinger's Equation in Configuration Space" describes the configuration-space N-particle and single-particle ($N = 1$) Schrödinger's equations. Calculation of probabilities using the configuration-space formulation is described in the section "Calculation of Probabilities Using the Configuration-Space Formulation". Finally, the possible factorization of the N-particle configuration-space wave function that occurs when the various particles are strictly non-interacting and statistically independent is discussed in the section "Factorization of Schrödinger's N-particle Configuration-Space Wave Function". Particles that do not interact and that have never interacted with each other, in that special case, may be considered statistically independent. It is noted in the section "Wave-Function Factorization or Not!", however, that in the more general case, particles are usually entangled, even when there is (presently) no interaction between them. In general, they are not independent. It is also noted that all charged particles in the world interact with each other, at least weakly, and are thus are always slightly entangled, with that entanglement exponentially growing in time, so that the special case never accurately applies.

---

[5] This appellation is offered by Bethe and Salpeter (1957).

The section "Born's Argument-Space Ambiguity" proceeds to examine some of the various discrepancies between the lab-space and configuration-space formulations of quantum theory. For example, one may note that the depictions in Figs. 3.1–3.7 are of quantum mechanical matter-wave propagation and particle propagation in lab space, i.e. propagation in the three-dimensional space in which we live. Unfortunately, the diagrams in Fig. 3.1 (from Born) and Fig. 3.2 (from Merzbacher), depicting the motion of waves in lab space around the nucleus of a single-electron hydrogen atom, are impossible to be drawn for a two-electron helium atom! (See the section "Born's Ambiguity's Misuse by the Lab-Space Formulation School" for observations and discussions of this fact by Dicke and Wittke (1960), Merzbacher (1961, 1970). Woops! Lab space doesn't get very far up the periodic table in describing the structure of atoms, does it? That is because a lab-space wave function, $\Psi_{lab}(\mathbf{r}_{lab})$, is limited to describing only a single-particle system. Also, the lack of any evident spatial dependence by a two-particle configuration-space wave function, $\Psi_{config,2}(\mathbf{r}_{1,config}, \mathbf{r}_{2,config})$, or by its associated probability density $|\Psi_{config,2}(\mathbf{r}_{1,config}, \mathbf{r}_{2,config})|^2$, prohibits these quantities from being considered as a valid description of a wave-like field propagating in lab space for $N > 1$ electron atoms, as per Figs. 3.1 and 3.2, whereupon $\Psi_{lab}(\mathbf{r}_{lab})$ does not and cannot describe waves propagating in lab space for helium.

The lack of ability for a lab-space wave function to describe $N > 1$ particle systems becomes a fatal difficulty for the lab-space formulation, especially when entanglement is required. Curiously, the importance of this observation seems heretofore to have gone unnoticed. It is shown in the section "Born's Argument-Space Ambiguity" that it apparently stems from a somewhat hidden ambiguity introduced by Born. In his textbook, he wrongfully pronounces the equivalence of configuration-space and lab-space wave functions. It is shown in the section "Born's Ambiguity's Misuse by the Lab-Space Formulation School" that there appears to be no rigorous method to allow a lab-space wave function and the associated lab-space formulation of quantum mechanics to be extended to describe $N \geq 2$ particle systems. Attempts by various books to demonstrate a direct connection between $\Psi_{lab}(\mathbf{r}_{lab})$ and $\Psi_{config,2}(\mathbf{r}_{1,config}, \mathbf{r}_{2,config})$ are examined in the section "Born's Ambiguity's Misuse by the Lab-Space Formulation School" and are found wanting. Thus, the wave functions are not equivalent, despite Born's pronouncement. This lack of equivalence is herein called *Born's argument-space ambiguity*. Correspondingly, there is no rigorous method evident to allow a lab-space wave function, along with the lab-space formulation also to describe entanglement! Born's important construct, the "conserved probability current" is examined in the section "Born's Conserved Probability Current as Re-Interpreted Using the Configuration-Space Formulation", wherein it is shown that it may be constructed only in lab space, and only makes sense in lab space. It thus may not be used to describe a pair of particles in an entangled state, e.g. as a pair of coupled currents.

Born's ambiguity is produced by a sneaky slight-of-hand. The ambiguity manifests itself by using the same ambiguous (multiply defined) symbol $\mathbf{r}$ to represent two very different quantities, $\mathbf{r}_{lab}$ and $\mathbf{r}_{1,config}$, with different meanings altogether of their arguments. Then, two different equations (Schrödinger's equation for $N = 1$ particle in lab space and in configuration space) that are formally the same in their appearance are produced and both are claimed to govern nature. Both equations use the common ambiguous symbol $\mathbf{r}$. Presto, since the equations formally appear to be the same, the equations are claimed to be equivalent, when in reality, they are not. The switch is done so seamlessly that no one is aware of the prestidigitation that has passed.

Details of the switch are outlined below in the section "Born's Ambiguity's Misuse by the Lab-Space Formulation School". Born is not alone in its use. It will then be seen in the section "Born's Argument-Space Ambiguity" that many of the quoted authors, in addition to Born, appear to have been ambiguous in their choice of propagation space. Messiah (1961) usually reminds the author when configuration space is being used, (but not always). He fails to do so in his discussions of scattering and quantum field theory. In his Chap. 6, *Classical Approximation and the WKB method*, he follows exactly this prescribed prestidigitation in his §4, *Classical Limit of the Schrödinger Equation* (pp. 222–228).

Born's notational ambiguity is revealed (and avoided) in the present article by simply and "inelegantly" displaying the different meanings of the symbols as they are used. The equations introduced in the sections "The Lab-Space Formulation of Quantum Mechanics" and "The Configuration-Space Formulation of Quantum Mechanics" are all taken verbatim from the above-mentioned quantum mechanics textbooks. However, they differ slightly and conspicuously in notation from those in the textbooks by the addition here of "Lab" and "config" subscripts. These obsequiously conspicuous "misquotations" are meant as a necessary clarification of these quotations. The subscripts are unceremoniously added to all symbols that describe a spatial and/or spatially dependent variable or operator. (Sorry for the necessary annoyance and loss of "elegance".) The additions extend to important symbols that are quoted from the various textbooks to indicate which space it is assumed that the quoted author is using, given that there are usually at least two possibilities to choose.

In the present article, in order to highlight Born's argument-space ambiguity, the subscript "config" is added to any dynamical variable or operator defined in a general configuration space, and the subscript "j, config" is added to a dynamical variable or operator defined for the j-th particle in an N-particle configuration space. Sometimes, where there still may be confusion (as in the section "Wave-Function Factorization or Not!"), the subscripts "particle1", "particle2", etc. are used to specify the setting of a specific particle's number index to a dummy index's value. Dependent variables that depend upon such dynamical variables, and/or upon operators in an N particle system, are given the subscript "config, N". This notation is used to prevent confusion of these variables with formally similar variables that are defined in lab space.

The sections "Quantum Field Theory 1—Quantization of Known Classical Fields" through "Quantum Field Theory 2—Second Quantization of Wave-Functions" examine quantum field theory. Standard quantum mechanics (via Born's conserved probability current) requires that the number of particles described by it to be constant. In nature, however, particles are created and annihilated by various processes, and these processes are not part of standard quantum mechanics. Standard quantum mechanics thus needs to be extended to account for a varying number of particles and to calculate how that number changes with time. Additionally, Einstein (1917) demonstrated an important need for what he called "*directional radiation bundles.*" These must exist as a new quantum mechanical component part of the electromagnetic field in order for the second law of thermodynamics to hold. These particle-like *directional radiation bundles,* now known as photons, are not present in the classical description of the electromagnetic field in terms of Maxwell's equations. A quantum mechanical modification thus needs to be added to the classical description of the field. Quantum field theory was correspondingly developed to handle these different evident requirements associated with photons, and with the creation and annihilation of massive particles. The first of these two different modifications to standard quantum mechanics is called Quantum Field Theory 1". It describes the quantization of known classical fields (light and sound) in the sections "Quantum Field Theory 1—Quantization of Known Classical Fields" through "Some Observations Regarding Which School Is "Proper"". The second, described in the section "Quantum Field Theory 2—Second Quantization of Wave-Functions", is called Quantum Field Theory 2. It describes the "second quantization" of matter-wave fields.

Fermi's (1932) formulation of the quantum theory of radiation is discussed in the section "Quantum Theory of Radiation and Quantum Electrodynamics". The relation of its quantized electromagnetic field to Einstein's (1917) need for *directional radiation bundles* is discussed in the section "Quantum Field Theory 1—Quantization of Known Classical Fields". It is noteworthy that Fermi (1932) tried to show that his quantized electromagnetic field demonstrated a causal behavior in real space–time. The section "von Neumann's Collapse of the Entangled Two-Photon Quantized Electromagnetic Field", however, shows that this aim cannot be achieved, because of the non-causal non-unitary evolution that it must undergo as required by von Neumann's collapse process. Said non-causal behavior is to be expected, and is predicted by experimental tests of Local Realism.

A scrutiny of Fermi's treatment reveals the existence of fields of two types. These two types are sometimes confused with each other via Born's ambiguity by others, but not by Fermi. Indeed, they are the same two types identified in standard quantum mechanics, i.e. they differ by their choice of argument space—lab space or configuration space. Things get worse. In Quantum Field Theory 2, various authors attempt to second quantize at least seven kinds of fields. It is thus observed in the section "Quantum Field Theory 2—Second Quantization of Wave-Functions" that second quantization of the matter-wave field evidently cannot proceed without a liberal use of Born's ambiguity.

Inspired by Bell's (1964) paper[6] and following the associated proposed experimental testing of local hidden variable theories by Clauser et al. (1969), Clauser and Horne (1974), added yet a fourth candidate school of thought for describing natural phenomena. Clauser and Horne (CH) originally named their formulation "Objective Local Theories". Clauser and Shimony (1978) reviewed it and renamed it "Local Realism".

---

[6] Bell (1964), in turn, was inspired by his reanalysis of Einstein, Podolsky, and Rosen (1935).

Bell–Clauser–Horne–Shimony Local Realism is formulated in lab space and provides experimental predictions that differ from those made by all of the above schools of quantum mechanics. Importantly, that feature allows Local Realism to be distinguished experimentally from those three schools of thought.

Local Realism's most attractive heuristic feature is that it provides very general lab space formulation for all theories of natural phenomena that attempt to describe real stuff in a real space–time framework, consistently with special relativity. It's most disappointing but also important heuristic feature is that it is soundly refuted by experiment. Local Realism describes tangible stuff, stuff that is present locally in space–time, stuff that can be put in a box, and stuff that can be used for storing bits of information. It thus provides a general framework for the space–time description of this stuff. An important requirement for the description is that it does not allow communication among any of the stuff to occur at super-luminal velocities, so as to somehow influence the results of experiments. Thus, Local Realism is consistent with special relativity. As an extension of classical field theories, Local Realism further allows non-deterministic evolution of the stuff that it describes, and has no required or seemingly artificial or presumed limitations to the precision of measuring devices.

Local Realism carefully defines a class of theories that attempt to describe *"real stuff in real space–time"*. By providing this definition along with its experimental predictions, Local Realism provides heuristic value in showing what quantum mechanics is not! Importantly, Local Realism provides experimental predictions that must be obeyed by any theory that attempts to describe *"real stuff in real space–time"*. Also importantly, the Clauser–Horne inequality's predictions differ from those made by quantum mechanics. Starting in 1972 with the first experiment by Freedman and Clauser (1972), followed by the second one by Clauser (1976), and then followed by a long list of confirming experimental refinements, Local Realism has been soundly refuted by experiment. Clauser (2017) reviews a partial list of twenty experiments that have been performed at many different laboratories around the world during the period 1972 through 2013 to test Local Realism. All but one of these experiments refute Local Realism's predictions. The theories basic predictions along with acid tests of its prohibition of internal super-luminal communication have all been now tested experimentally. Thus, quantum mechanics does not describe *real stuff in real space–time*.

Curiously, the lab space formulation of quantum mechanics bears uncanny and disturbing similarities to Local Realism. The difficulties experienced by the lab-space formulation are thus to be expected, given that the lab space formulation attempts to describe Local Realism's *real stuff in real space–time*. Indeed, the lab space formulation of quantum mechanics actually qualifies as a theory of Local Realism. It slickly avoids experimental refutation similarly to Local Realism's refutation, because it is limited to describing only single particle systems. Such systems, in turn, cannot exhibit non-local entanglement. Correspondingly, the lab space formulation of quantum mechanics is then incapable of making experimental predictions that can be tested by these experiments. Nonetheless, that deficiency should hardly be considered a salvation of the lab-space quantum mechanics formulation's viability.

The article's conclusions are presented in the final section.

Laboratory Space and Classical Fields

In this section we define what we refer to as "lab space" and "classical fields". While these definitions may seem obvious, tedious, and perhaps even boring, it is important to clarify them before proceeding, since they are frequently blurred[7] by practitioners of quantum theory and mathematical physics.

Laboratory space or lab space, for short, is the three-dimensional space in which we live, and the space in which Euclidean geometry is understood. It is the space used by geometrical vectors and by classical vector and scalar fields. Every point within one's laboratory has a unique position in what is called lab space. Said point is depicted by the symbol $\mathbf{r}_{\text{Lab}}$.

Geometric Vectors Are Defined in Lab Space

A classical geometrical vector (within in a classical vector field) is typically depicted as an arrow. It is a quantity with both a magnitude and a direction, and is commonly used to represent force and velocity. It is sometimes referred to as a *"vector of physics"*. Its definition does not require the existence of a coordinate system, and physical laws are often expressed in vector notation without reference to a coordinate system (see Margenau and Murphy (1943, 1956, p. 139)). It is often quantified by specifying its real scalar Cartesian components as projections on some set of spatial axes. When a classical geometrical vector is specified by its components, doing so thus requires a simultaneously defined coordinate system that has three real-valued components (exactly equal in number to the number of dimensions of lab space). In a classical vector field, a geometrical vector or "vector of physics" is anchored at every point in lab space, its three components are all real valued.[8]

Given a unique arbitrarily chosen point in the lab that is called the origin, every point in the lab may be located relative to the origin via the use of an appropriate Cartesian coordinate system. Each point in the lab also may be located by using a "position" vector, $\mathbf{r}_{\text{Lab}}$, that extends from the origin to it. Given a set of lab coordinates, every point in the laboratory thus has an associated vector position specified by

$$\mathbf{r}_{\text{Lab}} = \hat{\mathbf{e}}_x x_{\text{Lab}} + \hat{\mathbf{e}}_y y_{\text{Lab}} + \hat{\mathbf{e}}_z z_{\text{Lab}}, \tag{3.1}$$

where $\hat{\mathbf{e}}_x$, $\hat{\mathbf{e}}_y$, and $\hat{\mathbf{e}}_z$ are three orthogonal Cartesian unit basis vectors extending from the origin. Note that lab space necessarily has three, and only three, spatial dimensions. As an alternative to a Cartesian coordinate system, a variety of other 3D

---

[7] Said blurring is commonly called "interpreting".

[8] The term "real" here means having no imaginary component.

coordinate systems may be defined within said space without modifying in any way the geometrical properties of the space, itself.

Classical Fields Are Defined in Lab Space

A classical field is typically represented as a mathematical function, $f(\mathbf{r}_{Lab})$, of position, $\mathbf{r}_{Lab}$. The function assigns a unique value, f, (or values) to every point in the lab, i.e. the function's dependent variable, f, specifies the field's value at the point, $\mathbf{r}_{Lab}$. Many different fields may exist simultaneously. The value(s), f, also may take many forms—scalar, vector, tensor, set of scalars, set of vectors, or even more general mathematical forms. That is, f may take any basic generalized-spaghetti form that might be needed to specify the field's property or properties. If needed, it may be an n-tuple of numbers that are all defined simultaneously at the point, $\mathbf{r}_{Lab}$. The function's independent variable, or argument, is here represented by the dummy variable, $\mathbf{r}_{Lab}$. It specifies the position in the lab where the function's value or values apply Classical fields may be time varying, whereupon a second argument, t, is used to specify the time at which the function's assignment applies. In general, lab space is required for the description of the mechanics of continua, such as fluids, when no particles are evident or even present.

Classical Fields Are Used to Specify How Classical Stuff is Distributed Throughout Lab Space

The assumption that nature consists of *real stuff* distributed throughout lab space is the fundamental basis of both classical physics and Local Realism. Classical fields are used by classical physics and Local Realism to describe how the properties of *real stuff* are distributed throughout lab space. So-called "classical physics" (definitions vary) then provides a theory (usually, but not necessarily deterministic) that is a subset of Local Realism. Local Realism will be described in more detail in section "Bell–Clauser–Horne–Shimony Local Realism". It allows for a non-deterministic evolution of stuff. As a preface to that section, it should be noted that all of the expressions used by Local Realism qualify as "classical fields" and are functions of $\mathbf{r}_{Lab}$. The real-valued probability density, $|\Psi_{lab}(\mathbf{r}_{lab}, t)|^2$, also qualifies as a classical field defined in lab space.

Familiar examples of classical vector fields include the velocity field of a fluid, $\mathbf{v}(\mathbf{r}_{Lab}, t)$, and the electric field $\mathbf{E}_{Lab}(\mathbf{r}_{Lab}, t)$ from electromagnetic theory. The velocity field's vector may be decomposed with its associated Cartesian components defined as

$$\mathbf{v}_{Lab} = \hat{\mathbf{e}}_x v_{x,Lab} + \hat{\mathbf{e}}_y v_{y,Lab} + \hat{\mathbf{e}}_z v_{z,Lab}. \tag{3.2}$$

Given some quantity of the stuff that is described by a field whose value varies as a function of position within the lab, there is an important differential vector operator that operates on the quantity's value (only) in lab space. It is the lab-space gradient operator:

$$\nabla_{Lab} = \hat{e}_x \partial/\partial x_{Lab} + \hat{e}_y \partial/\partial y_{Lab} + \hat{e}_z \partial/\partial z_{Lab}. \tag{3.3}$$

It may be used to calculate the rate of change of a function's value with respect to spatial position within the lab. Important theorems of vector analysis (Gauss's theorem, Green's theorem(s), and Stokes theorem, etc.) all apply rigorously to scalar and vector fields that are defined in lab space. Waves in linear classical fields propagate in lab space, typically governed by the classical wave equation. In such cases, the time dependence of the equation can be factored out, and the result is the time-independent linear Helmholtz equation.

Lab Space is not the Same as a General Vector Space

At this point in the discussion, it is worth commenting on what lab space is not. Unfortunately, some authors have imprecisely adopted the use of the words "vector space" and "vector" to describe abstract mathematical constructs in quantum mechanics that may or may not include "lab space" and "classical geometrical vector" in lab space.

A "geometrical vector" is sometimes confused with an "algebraic vector". The two types of vectors are sometimes treated as equivalent interchangeable entities. They are not! Apostol (1961, V1, p. 252) discusses *"another approach to vector algebra called the abstract or axiomatic approach, ... Instead, vectors and vector operations are thought of as undefined concepts of which we know nothing except that they satisfy a certain set of axioms. Such an algebraic system, with appropriate axioms is called a linear space or a linear vector space. ..."* An algebraic vector (within a vector space) is commonly represented as an *"ordered n-tuple"* of numbers. The number, n, of numbers in the n-tuple may be greater than three, even infinite, and the numbers themselves may be complex.

An ordered n-tuple of numbers is not, in general, associated with a specific position in lab space. It does not necessarily qualify as a geometrical vector, especially when n is greater than three, and especially when the numbers themselves are complex. For the purposes of the present discussion, a "vector space" is taken here to mean a set of numbers (an algebraic vector) that somehow functionally depends on some other set of numbers, and nothing more.

The lab-space gradient operator defined by (3.3), when operating on a lab-space classical scalar field produces a lab-space vector field with a unique geometric vector thereby defined at every point in lab space. On the other hand, an algebraic vector defined in configuration space, (in the section "The Configuration-Space Formulation of Quantum Mechanics") is not a lab-space geometric vector. As a preface to that section, it should be noted that the real-valued probability "density",

$|\Psi_{config,N}(\mathbf{r}_{1,config}, \mathbf{r}_{2,config}, ..., \mathbf{r}_{N,config}, t)|^2$ correspondingly does not qualify as a classical field. Indeed, it is also pointed out in that section that Messiah (1961, p. 164) succinctly defines *"The wave functions of wave mechanics are the square integrable functions of configuration space."* It should also be stressed that the special case with $N = 1$ does not change the space on which $\Psi_{config,N}$ depends from configuration to lab.

Bell–Clauser–Horne–Shimony Local Realism

The term "Local Realism" describes a large class of theories, first explicitly defined by Clauser and Horne (CH) in 1974. Clauser and Horne named the class, *"Objective Local Theories"*. Significant contributions were made to the theory by Bell and Shimony. In response to Clauser and Horne (1974), Bell offered a different (unpublished) version which he called *"The Theory of Local Beables"*. Local Realism was refined and clarified jointly by Bell et al. (1976–1977), regarding methods used to specify the apparatus parameters and differences between *"Objective Local Theories"* and *"The Theory of Local Beables"*. Local Realism theory was further refined in the review article by Clauser and Shimony (1978), who gave the class of theories the name Local Realism.

The Locality principle is based on special relativity. It asserts that nature does not allow the propagation of information faster than light to thereby influence the results of experiments. Without Locality, one must contend with paradoxical causal loops, as are now popular in science fiction thrillers involving time travel. Upholding Locality is effectively denying the possible reality of causal loops.

Realism is a philosophical view, according to which external reality is assumed to exist and have definite properties, whether or not they are observed by someone. Another way of describing what is meant by Realism is to say that it specifies that nature consists of "objects", i.e. of stuff that is distributed throughout space–time with "objective reality". Realism assumes that stuff, i.e. objects, with spatial positions and structure, exist and have inherent properties on their own. It does not require that these properties fully determine the results of an experiment locally performed on said stuff. Instead, in a possibly non-deterministic world, it simply allows the properties of stuff (a discrete object, or objects, or continuum objects) to somehow influence the probabilities of experiments being performed locally on it. There is also nothing in this specification that prohibits an act of observation or measurement of an object or property from influencing, perturbing and/or even destroying said properties of the object or royally messing up the property.[9] Realism thus assumes that an object's properties determine minimally the probabilities of the results of experiments locally

---

[9] Whether or not there is a disturbance of a property (or properties) that is made during its measurement certainly does not necessarily mean that there was not a well-defined property of the stuff existing prior to a measurement. Such a disturbance, if present, simply indicates a clumsy measuring apparatus or procedure. At worst, such a clumsy measurement only messes up the properties available for a subsequent measurement. Also, simply because no one at present can think of how to build

performed on it. Realism, under the additional constraint of locality, become Local Realism. Local Realism assumes that the results of said experiments do not depend on other actions performed far away, especially when those actions are performed outside of the light-cone of the local experiment.

Local Realism is the combination of the philosophy of realism with the principle of locality. Local Realism describes "real stuff in real space–time". Stuff, objects, and their associated properties, as referred to here, are what John Bell called Local Beables, and what Einstein Podolsky and Rosen (1935) called "elements of reality". The properties of an object constitute a description of the stuff that is "really there" in nature, independently of our observation of it. When we perform a "measurement" of these properties, we don't really need to know what we are actually doing, or what we are really measuring. The fundamental assumption underlying Local Realism that what is "really there", even if we don't know exactly what it is, nonetheless somehow influences what we observe, even if said influence is inherently stochastic and/or perhaps irreproducible from one measurement to the next. Recall that Einstein et al. (1935) attempted to define an object's properties as something that one can measure, but they further required that the measurement result be predictable with certainty. However, given Ben Franklin's observation that the only predictions that are certain in life are for death and taxes, said definition becomes meaningless, because it describes nothing that can ever be present in reality. Local Realism's definition is much looser and requires no predictions with certainty.

Precisely how did Clauser and Horne (1974) define an object with such extreme generality? For the purposes of Local Realism and its tests via Bell's Theorem and the Clauser–Horne (CH) inequality, a purely operational definition of an object suffices. An object (or collection of objects) is stuff with properties that one can put inside a "box", wherein one can then perform measurements inside said box and get results whose values are presumably influenced by the object's properties.[10] What is a "box"? A box is defined as a closed Gaussian surface,[11] inside of which one can perform said measurements of said properties. Three dimensional (temporally shrinking) Gaussian Surface "boxes" are shown in Fig. 3.8. For Local Realism, such a box becomes a four-dimensional Gaussian surface consisting of the backward light cone (extending to $t = -\infty$) enveloping a three dimensional box, that contains the object(s) being measured, at the time that they are being measured. Four dimensional Gaussian Surface "boxes" are depicted in Fig. 3.9.

Familiar examples of "classical" objects that can be put into boxes are solar systems, airplanes, shoes, trapped clouds of atoms, single trapped atoms, electrons, y-polarized photons, a single bit of information, etc. All of these can be put into a sufficiently large box and have their properties (e.g. color, mass, charge, etc.)

---

an improved measuring apparatus that will allow a super-observer to make such a measurement also does not preclude the existence of specific properties characteristic of each point in space.

[10] See especially, Clauser and Horne (1974, Footnotes 10–15).

[11] Gauss showed that a "Gaussian surface" (sometimes called a closed surface) is one that divides all of space into two disjoint volumes, wherein one of these volumes may be called the inside, and the other the outside.

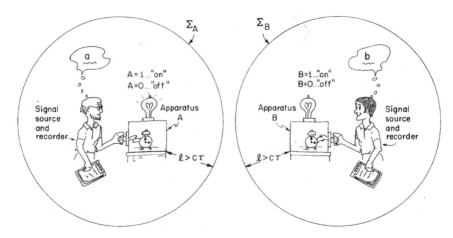

Fig. 3.8 Snapshot of a pair of 3D (temporarily shrinking) Gaussian-surface boxes labeled $\sum_A$ and $\sum_B$, that envelop two separated apparatuses. Each apparatus is measuring a selected property, of an object, giving the binary result, count (on) or no-count (off). The associated property is selected by its associated parameter setting, **a** or **b**. The boxes are space-like separated. Signal sources in each generate the apparatus parameter settings, randomly. In this example, they are generated via human "free-will", with both settings assumed to be independently randomly chosen and not influenced by their communal past

measured. Or can they? Via Bell's Theorem experiments, one may ask—are there examples of objects that cannot be put inside such boxes[12]? If so, such objects cannot be described by Local Realism. Furthermore, if there are parts of nature that cannot be described by Local Realism, then Local Realism must be discarded as a description of all of nature. Sadly, (for Local Realism advocates) experiments now show that the individual particles comprising a quantum-mechanically entangled pair of particles are parts of nature that cannot be described by Local Realism.

Figure 3.8 shows a disjoint, space-like-separated pair of 3D Gaussian-surface "boxes" labeled $\sum_A$ and $\sum_B$.[13] Each box surrounds a pair of objects, each being measured respectively by a pair of measuring apparatuses. Figure 3.9 displays these same two boxes evolving in 2D x-t space. The boxes $\sum_A$ and $\sum_B$, are shrinking at the speed of light as time progresses forward to their positions shown in Fig. 3.8. Two 4D Gaussian-surface boxes, are thus each formed by the backward light cones that envelop 3D boxes labeled $\sum_A$ and $\sum_B$. The 4D boxes contain associated 4D volumes, Γ_A and Γ_B. Figure 3.9 thus shows a (2D projection) space–time diagram of the stuff in space–time, that is contained within these 4D volumes, and that can influence the probability of a count or no-count at each apparatus.

---

[12] The fact that the simplest possible object—a single bit of information—cannot be put into a "box", in turn gives rise to the field of quantum information.

[13] Figure 3.8 was first presented by the author at the 1976 International"Ettore Majoranna" Conference in Erice, Sicily on "Experimental Quantum Mechanics. The conference was organized by John Bell, Bernard d'Espagnat, and Antonino Zichichi. With present-day jargon, the characters labeled "Signal source and recorder" would now be named Bob and Alice.

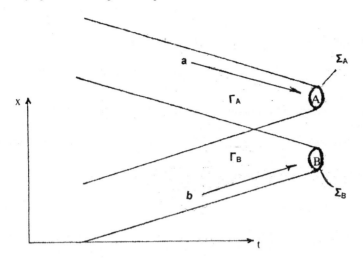

Fig. 3.9 Space–time diagram of stuff in lab space described by Local Realism. Two backward-light-cone 4D Gaussian-surfaces containing associated volumes, Γ_A and Γ_B, respectively, envelope the 3D boxes, $\sum_A$ and $\sum_B$, shown in Fig. 3.8. Objective state properties $\lambda(\mathbf{r}_{lab},t)$ are defined at every space–time point $(\mathbf{r}_{lab},t)$. The probability of a count at apparatus A or B may depend only on properties, $\lambda(\mathbf{r}_{lab},t)$, contained within the associated 4D volume Γ_A or Γ_B. An alternative parameter-selection method is used here with respect to that used in Fig. 3.8. Apparatus parameter settings, **a** and **b**, are now set by signals from events occurring outside of the overlap region of Γ_A and Γ_B. A recent experiment performed by Kaiser and Zeilinger satisfies this condition by having the choices for **a** and **b** set by light from two distant quasars (see Public Broadcasting System, (2018) "*Einstein's Quantum Riddle*", *Nova* television documentary.)

Let us use the symbol, $\lambda(\mathbf{r}_{lab}, t)$, to represent the complete set of properties at any point in space–time, $(\mathbf{r}_{lab}, t)$. It may have whatever level of complexity that is necessary to do so. The properties are assumed to be randomly distributed with an ensemble probability density $\rho(\lambda(\mathbf{r}_{lab}, t))$. Given the arrangement shown in Figs. 3.8 and 3.9 and the assumptions made by Local Realism, the properties $\lambda(\mathbf{r}_{lab}, t)$ located at a point $(\mathbf{r}_{lab}, t)$ that is within Γ_A <u>may</u> influence the probability of a count at apparatus A. Define the incremental probability of a count at apparatus A when the apparatus is configured with parameter setting, **a**, to be $p_A(\mathbf{a}, \lambda(\mathbf{r}_{lab}, t))$. This definition holds, independently of what happens at apparatus B. The corresponding incremental probability of a count at apparatus B with parameter setting, **b**, independently of what happens at detector A, is defined to be $p_B(\mathbf{b}, \lambda(\mathbf{r}_{lab}, t))$, for properties, $\lambda(\mathbf{r}_{lab}, t)$, located at a point, $(\mathbf{r}_{lab}, t)$, within Γ_B. The integrated probability, $p_A(\mathbf{a})$, of a count at apparatus A, irrespective of what happens at apparatus B, that is due the influence of all properties distributed throughout the 4D volume Γ_A is then given by

$$p_A(\mathbf{a}) = \int_{\Gamma_A} d\mathbf{r}_{lab}\, dt\, p_A(\mathbf{a}, \lambda(\mathbf{r}_{lab}, t)) \rho(\lambda(\mathbf{r}_{lab}, t), \mathbf{r}_{lab}), \qquad (3.4)$$

and the probability of a count, $p_B(\mathbf{b})$, at apparatus B, that is due the influence of all properties distributed throughout the 4D volume Γ_B is given by

$$p_B(\mathbf{b}) = \int_{\Gamma_B} d\mathbf{r}_{lab} \, dt \, p_B(\mathbf{b}, \lambda(\mathbf{r}_{lab}, t)) \, \rho(\lambda(\mathbf{r}_{lab}, t), \mathbf{r}_{lab}). \tag{3.5}$$

Except for correlations caused by common causes in the overlap region, the results at A and B are otherwise independent. The parameter settings $\mathbf{a}$ and $\mathbf{b}$ are also chosen independently. Locality then requires that the incremental joint probability of a "coincident" count at both detectors A and B, caused by properties located at a point, $\lambda(\mathbf{r}_{lab}, t)$, is then given by

$$p_{AB}(\mathbf{a}, \mathbf{b}, \lambda(\mathbf{r}_{lab}, t)) = p_A(\mathbf{a}, \lambda(\mathbf{r}_{lab}, t)) \, p_B(\mathbf{b}, \lambda(\mathbf{r}_{lab}, t)). \tag{3.6}$$

The factored form, (3.6), is the key ingredient for the CH argument to proceed. The integrated joint probability, $p_{AB}(\mathbf{a}, \mathbf{b})$, of a "coincident" count at detectors A and B for properties located at a space–time point <u>anywhere</u> that might influence the joint result is then

$$\begin{aligned} p_{AB}(\mathbf{a}, \mathbf{b}) &= \int_{\Gamma_A \cup \Gamma_B} d\mathbf{r}_{lab} \, dt \, p_{AB}(\mathbf{a}, \mathbf{b}, \lambda(\mathbf{r}_{lab}, t)) \, \rho(\lambda(\mathbf{r}_{lab}, t), \mathbf{r}_{lab}) \\ &= \int_{\Gamma_A \cup \Gamma_B} d\mathbf{r}_{lab} \, dt \, p_A(\mathbf{a}, \lambda(\mathbf{r}_{lab}, t)) \, p_B(\mathbf{b}, \lambda(\mathbf{r}_{lab}, t)) \, \rho(\lambda(\mathbf{r}_{lab}, t), \mathbf{r}_{lab}). \end{aligned} \tag{3.7}$$

Equations (3.5)–(3.7) were used by Clauser and Horne (1974), who found them sufficient to derive the CH inequality as an experimental prediction by all theories in the class defined as Local Realism for the results of an experiment shown with the configuration of Figs. 3.8 and 3.9. The CH inequality is

$$\begin{aligned} -1 \leq p_{AB}(\mathbf{a}, \mathbf{b}) &- p_{AB}(\mathbf{a}, \mathbf{b}') + p_{AB}(\mathbf{a}', \mathbf{b}) + p_{AB}(\mathbf{a}', \mathbf{b}') \\ &- p_A(\mathbf{a}') - p_B(\mathbf{b}) \leq 0. \end{aligned} \tag{3.8}$$

An earlier version of (3.8), the CHSH inequality, was first given by Clauser et al. (1969), for a more restrictive deterministic class of theories. The CH inequality, (3.8) reduces to the CHSH inequality[14] when various modifications are made for the experimental arrangement (the inclusion of source heralding, or Clauser and Horne's no-enhancement assumption) so that the CHSH inequality can also be used for experimental tests of Local Realism.

By carefully defining what "Local Realism" is, Clauser and Horne provide heuristic value in showing what quantum mechanics is not. Local Realism provides experimental predictions that must be obeyed by any theory that attempts to describe *"real stuff in real space–time"*. Importantly, the CH prediction (3.8) differs from

---

[14] Both the CH and CHSH inequalities are examples of what Clauser and Horne (1974) named "Bell Inequalities". See Clauser (2017) for a review of the various testable and tested Bell inequalities.

that made by quantum mechanics. Experimental tests for either the CHSH or the CH inequality are difficult, and it has taken many years for technology to advance to the point where loophole-free direct tests of (3.8) can be performed. In an effort to allow testing with 1970's technology, Clauser and Horne (1974) provided a very plausible auxiliary assumption (their no-enhancement assumption), which allowed the technology of that era to provide an experimental test. Clauser (2017) provides a review of this and other assumptions that have been made to allow experimental testing of Local Realism over the years, as technology for performing such tests has improved.

Starting in 1972 with the first experiment by Freedman and Clauser (1972), followed by the second one by Clauser (1976), and then followed by a long list of confirming experimental refinements continuing to the present, Local Realism has been finally conclusively refuted by experiment. Clauser (2017) provides a partial list of twenty experiments that have been performed at different laboratories around the world during the period 1972 through 2013 to test Local Realism. All but one of these experiments refute Local Realism's predictions. The theory's basic predictions along with "acid tests" of its prohibition of internal super-luminal communication have all been now tested experimentally. Clauser and Shimony's (1978) conclusion regarding Local Realism still stands—"Consequently, it can now be asserted with high confidence that either the thesis of Realism or that of locality (or perhaps even both) must be abandoned. Additionally, any theory that falls within the scope of its underlying assumptions must also be abandoned."

As we shall demonstrate below, the lab-space formulation of quantum must similarly be abandoned.

The Lab-Space Formulation of Quantum Mechanics

Textbooks that promote the lab-space formulation of quantum mechanics generally employ conceptual models that are motivated and justified by the concepts depicted by Figs. 3.1–3.7. Once Schrödinger's equation has been formulated in lab space using this conceptual model, these books then immediately proceed to use it, lab space, and the conceptual model to solve a long list of single-particle problems. The problems typically include free-particle motion, motion of a single particle confined in a 1D and 3D square well, motion of a single particle at a potential step and through a potential barrier, finding the energy levels of a hydrogen atom, quantized orbital angular momentum, the scattering of a single particle by a central potential, and semi-classical radiation theory.

Max Born appears to have been a father of the lab-space formulation of quantum mechanics. It is described in detail in his seminal book, *Moderne Physik*, Born (1933, 1935, 1969). Subsequently authored books by Schiff (1955), Dicke and Wittke (1960), Eisberg (1961, 1967), Merzbacher (1961, 1970), Feynman (1948), Feynman and Hibbs (1965), Eisle (1964), and French and Taylor (1978), all formulate quantum mechanics in lab space, whether they mention this fact or not.

A basic tenant of the lab-space formulation is that Schrödinger's wave function is treated as a "classical field" that is defined and formulated in lab space, as per section "Laboratory Space and Classical Fields". Textbooks that use a lab-space formulation treat Schrödinger's wave function, $\Psi_{Lab}(\mathbf{r}_{Lab}, t)$, as a field that propagates in lab space, as depicted in Figs. 3.1–3.7. Correspondingly, the real-valued probability density, $|\Psi_{lab}(\mathbf{r}_{lab}, t)|^2$, also propagates in lab space as a classical field, like one found in electrodynamics or fluid mechanics, except that it is subject to the somewhat mysterious "statistical interpretation". That "interpretation" is universally attributed to Max Born.

Born, in his textbook (see quote from his book in the section "Born's Argument-Space Ambiguity"), describes the *"wave amplitude"*, $\Psi_{lab}(\mathbf{r}_{lab}, t)$, as an *"ordinary physical magnitude"*. Born's modification of the traditional classical field concept occurs with his invention of the *"statistical interpretation"*. He uses $|\Psi_{lab}(\mathbf{r}_{lab}, t)|^2$ to describe, not only probability density in lab space, but also uses it to describe particle density as a function of lab-space position, wherein particles move at the group velocity of a wave that propagates in lab space. (See the section "Born's Argument-Space Ambiguity" for a detailed discussion of Born's description.) Thus, he uses it to describe the "density" of stuff in lab space, although the term "particle density" is obviously difficult to define for a single-particle theory. His metaphoric usage is perhaps excused by his also calling it a "probability density".

Given the definition of lab space from the section "Laboratory Space and Classical Fields", and the definition of Local Realism from the section "Bell–Clauser–Horne–Shimony Local Realism", then $Re[\Psi_{lab}(\mathbf{r}_{lab}, t)]$, $Im[\Psi_{lab}(\mathbf{r}_{lab}, t)]$, and $|\Psi_{lab}(\mathbf{r}_{lab})|^2$ all qualify as classical scalar fields. Unfortunately, the lab-space formalism of quantum mechanics only applies to single-particle systems! It will be shown in the section "Born's Argument-Space Ambiguity" that there is no rigorous method to allow lab-space wave functions, along with a lab-space formulation of quantum mechanics, to be extended to describe $N \geq 2$ particle systems. This impossibility might be expected, since Local Realism's formulation and Bell's Theorem forbids it. Correspondingly, experiments that refute Local Realism also refute a lab space formulation of quantum mechanics.

Single Particle Schrödinger's Equation in Lab Space

Authors formulating Schrödinger's equation in lab space typically start by "searching" for a partial differential equation that is, itself, formulated in lab space. Schiff (1955, pp. 20–22), for example, formulates Schrödinger's equation as a direct analogy to sound waves. Merzbacher (1961, 1970, pp. 34–42), Eisberg (1961, 1967, pp. 166–170, Dicke and Wittke (1960, pp. 23–36), and others rely on conceptual aids like Figs. 3.1–3.7. The lab-space formulation generally starts with a requirement that its solution for a free particle be a plane wave, as per

$$\Psi_{Lab}(\mathbf{r}_{Lab}, t) = \exp(i(\mathbf{k} \cdot \mathbf{r}_{Lab} - \omega t)), \tag{3.9}$$

with its propagation vector is given by

$$|\mathbf{k}| \equiv 2\pi / \lambda_{\text{deBroglie}},$$

and where the fundamental relations of quantum theory, deBroglie's and Einsteins's relations,

$$\lambda_{\text{deBroglie}} \equiv h/|\mathbf{p}|, \ E = \hbar\omega, \mathbf{p} = \hbar\mathbf{k}. \tag{3.10}$$

also apply. Equation (3.9) then becomes

$$\Psi_{\text{Lab}}(\mathbf{r}_{\text{Lab}}, t) = \exp\left(i(\mathbf{p} \cdot \mathbf{r}_{\text{Lab}}/\hbar - E\, t/\hbar)\right). \tag{3.11}$$

The technique employed to obtain the desired partial differential equation is to use a somewhat mysterious but now standard "operator substitutions trick"[15] for these variables, as per

$$E = \hbar\omega = p^2/2m + V \rightarrow -i\hbar\, \partial/\partial t,$$
$$\mathbf{p} = \hbar\mathbf{k} \rightarrow -i\hbar\nabla_{\text{Lab}},$$
$$|\mathbf{p}|^2 \rightarrow -\hbar^2\nabla^2_{\text{Lab}}$$

The equation's component parts are then combined using the indicated operator-substitutions trick to yield the desired partial differential equation—Schrödinger's time dependent equation in lab space,

$$[-(\hbar^2/2m)\nabla^2_{\text{Lab}} + V]\,\Psi_{\text{Lab}}(\mathbf{r}_{\text{Lab}}, t) = i\hbar\, \partial/\partial t\, \Psi_{\text{Lab}}(\mathbf{r}_{\text{Lab}}, t). \tag{3.12}$$

Additionally, the so-called Hamiltonian operator is defined as

$$H_{\text{Lab}} \equiv -\hbar^2\nabla^2_{\text{Lab}} + V(\mathbf{r}_{\text{Lab}}), \tag{3.13}$$

giving a more compact form of Schrödinger's time dependent equation in lab space,

$$H_{\text{Lab}}\Psi_{\text{Lab}}(\mathbf{r}_{\text{Lab}}, t) = i\hbar\partial/\partial t\, \Psi_{\text{Lab}}(\mathbf{r}_{\text{Lab}}, t). \tag{3.14}$$

Stationary state solutions, $\Psi_{\text{Lab}}(\mathbf{r}_{\text{Lab}})$, are readily found by factoring out the time dependence using

$$\Psi_{\text{Lab}}(\mathbf{r}_{\text{Lab}}, t) = \exp\left(i(\mathbf{p} \cdot \mathbf{r}_{\text{Lab}}/\hbar - Et/\hbar)\right) = \Psi_{\text{Lab}}(\mathbf{r}_{\text{Lab}})\exp\left(-iEt/\hbar\right), \tag{3.15}$$

---

[15] This mysterious trick is referred to by Messiah (1961, p. 885) as the "*Schrödinger correspondence rule*".

whereupon Schrödinger's time-independent equation in lab space is given by

$$H_{Lab}\Psi_{Lab}(\mathbf{r}_{Lab}, t) = E\Psi_{Lab}(\mathbf{r}_{Lab}, t). \tag{3.16}$$

Equation (3.15) will be recognized as the time-independent Helmholtz equation in lab space, familiar from electromagnetic theory. A solution to (3.15) is then an eigenfunction of the operator H_{Lab}, with the eigenvalue E.

Born's Probability Density and Conserved Probability Current Defined in Lab Space

Born's statistical interpretation starts with his definition of the scalar field,

$$P(\mathbf{r}_{lab}, t) \equiv |\Psi_{lab}(\mathbf{r}_{lab}, t)|^2, \tag{3.17}$$

as the probability density in lab space. A common assertion by Born and echoed by all quantum mechanics books is that is that $P(\mathbf{r}_{lab}, t)$ describes a *"wave of probability"*. Under the lab space formulation and Born's "statistical interpretation", the probability for detecting the single particle's presence with a detector positioned at $\mathbf{r}_{lab} = \mathbf{r}_{det}$, within the differential volume element,

$$d^3\mathbf{r}_{lab} \equiv dx_{Lab}dy_{Lab}dz_{Lab},$$

between the times, t_{det} and $t_{det} + dt$, is given by

$$P(\mathbf{r}_{lab} = \mathbf{r}_{det}, t = t_{det}) \equiv |\Psi_{lab}(\mathbf{r}_{lab} = \mathbf{r}_{det}, t = t_{det})|^2. \tag{3.18}$$

Given Born's statistical interpretation, $P(\mathbf{r}_{lab}, t)$ also describes the local particle/probability density in lab-space position. The probability density's normalization is found by integrating the dummy variable, $\mathbf{r}_{lab}$, over all space, as per

$$1 = \int d^3\mathbf{r}_{lab} \, P(\mathbf{r}_{lab}, t). \tag{3.19}$$

For a wave function that is not square integrable, like that of a plane wave, a finite normalization volume alternatively may be used. Equation (3.19) specifies that the probability of finding the particle somewhere within the lab-space normalization volume is always unity.

In order to link his probability density to particle density, Born further assumes that the particle moves within the lab with a probability flux that he calls a "probability current". It is described by a classical vector field,[16] $\mathbf{S}(\mathbf{r}_{lab}, t)$. It is defined such that the particle and/or probability flux impinging on a detector's surface is given by $\int \mathbf{S} \cdot d\mathbf{A}$, where $d\mathbf{A}$ is the detector-surface normal's differential area, and where the integral extends across the detector's surface. Given the normalization (3.19), the appropriate flux-density vector is then just the single-particle's velocity vector[17] defined locally at the point, $\mathbf{r}_{lab}$, as

$$\begin{aligned} \mathbf{S}(\mathbf{r}_{lab}, t) &\equiv \mathrm{Re}[\Psi_{lab}^*(\mathbf{r}_{lab}, t)\,(\mathbf{p}/m)\Psi_{lab}(\mathbf{r}_{lab}, t) \\ &= \mathrm{Re}[\Psi_{lab}^*(\mathbf{r}_{lab}, t)\,(-i\hbar\nabla_{Lab}/m)\Psi_{lab}(\mathbf{r}_{lab}, t)]. \end{aligned} \qquad (3.20)$$

The second line of (3.20) uses the "standard" operator substitutions trick used above in the section "Single particle Schrödinger's equation in lab space". To show consistency among the definitions, Born relies on the fact that $P(\mathbf{r}_{lab})$, $\mathbf{S}(\mathbf{r}_{lab}, t)$, $\mathrm{Re}[\Psi_{lab}(\mathbf{r}_{lab}, t)]$, and $\mathrm{Im}[\Psi_{lab}(\mathbf{r}_{lab}, t)]$ are all classical scalar and vector fields defined in lab space, as per the definitions given above in the section "Laboratory Space and Classical Fields". Green's theorem then applies to these fields. By using Green's theorem and Schrödinger's equation (3.12) together, Born proceeds to show that the fluid-flow conservation equation,

$$\partial/\partial t\, P(\mathbf{r}_{lab}, t) + \nabla_{Lab} \cdot \mathbf{S}(\mathbf{r}_{lab}, t) = 0, \qquad (3.21)$$

applies to his probability (and particle) density and flux. Given the conservation equation (3.21), he then refers to the vector field, $\mathbf{S}(\mathbf{r}_{lab}, t)$, as a "conserved probability current". Born (1933, 1935, 1969, Appendix XX) then demonstrates the use of (3.21) by applying it to the problem of flux conservation in Rutherford scattering. It is important to notice that equation (3.21) intimately relies on the use of the lab-space gradient operator, defined above by (3.3).

Born's presentation and definitions became an immediate hit and were universally adopted. They are repeated in many textbooks, regardless of the book's choice of propagation space. Textbooks that give this presentation in lab space include Schiff[18] (1955, pp. 22–24), Dicke and Wittke (1960, pp. 60–62), Merzbacher (1961, 1970, pp. 35–37), Eisberg, (1961, 1967, pp. 172–175) and French and Taylor (1978,

---

[16] Schiff (1955) and Dicke and Wittke (1960) use the symbol $\mathbf{S}$ for the probability current density, while Messiah (1961) uses the symbol $\mathbf{J}$.

[17] Dicke and Wittke (1960, pp. 60–62) say, "One may think of this wave as representing a swarm of particles with an average density of one particle per cubic centimeter. In this case, the particles are moving with momentum $m\mathbf{v}$, or have a velocity $\mathbf{v} = \mathbf{p}/m$. With this velocity and with an average density of one particle per cubic centimeter, v particles per second pass through a surface area of one square centimeter perpendicular to the direction of motion of the particles. This constitutes the probability flux of the wave."

[18] Schiff (p. 24) says "It is thus reasonable to interpret $\mathbf{S}(\mathbf{r}, t)$ given by Eq. (7.3) as a *probability current density*." Further on, he hedges and says "While this interpretation of $\mathbf{S}$ is suggestive, it must be realized that $\mathbf{S}$ is not susceptible to direct measurement in the sense in which P is. Nonetheless, it is sometimes helpful to think of $\mathbf{S}$ as a flux vector."

pp. 374–378). Textbooks that give the argument in configuration space are discussed in the section "Born's Argument-Space Ambiguity".

The Configuration-Space Formulation of Quantum Mechanics

The configuration-space school is described in John von Neumann's seminal textbook, *Mathematiche Grundlagen der Quantenmechanik*, (1932, 1955), and also in Messiah (1961). Von Neumann's (1932, 1955) textbook, unlike it's contemporary, Born's (1933, 1935, 1969) textbook, promotes the calculation of predictions for the results of experiments via the use of purely abstract mathematical tools with no tangible space–time counterparts. It is formulated using a very general abstract k-dimensional argument space for the wave function that is called configuration space. The quantum mechanical system being described by the school's formalism is very general, and the associated wave function, Ψ_{config}, is similarly very general. It describes a "system" with k-degrees of freedom. The associated Schrödinger's equation can be tailor-made to fit any system by specifying its degrees of freedom and the associated Hamiltonian. The wave function can describe a system comprised of any number, N, of particles, by letting the degrees of freedom be the positions of these N particles. The number of particles in the system may include the special cases $N = 1$ and $N = 0$. By including the $N = 0$ case, the existence of some finite number of particles is thus optional. The configuration-space formalism far surpasses the generality of the systems that can be described by those covered by the lab-space formalism. The latter is limited to describing only single-particle systems. It should be stressed that lab space is never used within the configuration-space formalism, except for specifying the locations of detectors at the end-point of a particle's presumed trajectory.

Configuration Space

The configuration-space school wave function's arguments are provided by configuration space. That space is an abstract mathematical vector space used for specifying the configuration of a general dynamical system. It originated with the Hamilton–Jacobi theory of classical mechanics, and is commonly used for describing the dynamics of a very general system with k degrees of freedom, and especially for describing systems where the forces acting between particles and their constraints are unknown.

Configuration space can be used in either classical mechanics or quantum mechanics. The system may consist of a continuous field (a continuum), such as a fluid's density, where there are no particles present. One way of doing so is to subdivide lab space into very small cells. The necessary degrees of freedom may then consist of the set of field values at each cell. Alternatively, the field's value may be decomposed as a sum of normal modes, and the degrees of freedom then taken to be the modes' expansion coefficients. Configuration space is also useful for describing the dynamics of a (non-continuum) system consisting of N particles. In that case, $k = 3$ N degrees of freedom are used to specify the set of N positions of the N particles. For classical mechanical systems, Goldstein (1950, p. 11) in his book *Classical Mechanics*, defines "degree of freedom" as an allowed unconstrained motion along an independent generalized coordinate. In quantum mechanics, configuration space is generalized beyond Goldstein's definition to include motion in energetically forbidden domains, thereby to allow barrier penetration and tunneling. It is also generalized to allow non-classical motions or other variations such as spin, etc. For example, Bjorken and Drell (1964, p. 2) use a configuration-space wave function, $\Psi_{config}(q_{1,config}, \ldots, q_{k,config}, s_{1,config}, \ldots, s_{k,config}; t)$, whose arguments are the system's degrees of freedom, and include the variables, $s_{1,config}, \ldots, s_{k,config}$, which are the non-classical spin degrees of freedom of the various particles in their system.

Configuration space is adapted for use by quantum mechanics by what is called "transformation theory". Configuration space is a subset of what is called phase space, which further includes the set of k-generalized momenta conjugate to the k-degrees of freedom. An important feature of quantum theory is that Ψ_{config} depends only this subset of variables, e.g. on either the k-degrees of freedom themselves, or on the k-generalized momenta conjugate to them, or on any possible linear transformation among these that provides some other transformed set of k-degrees of freedom.

Messiah (1961, p. 164) succinctly defines the wave function space used by the school by saying *"The wave functions of wave mechanics are the square integrable functions of configuration space, that is to say, the functions* $\Psi_{config}(q_1, \ldots, q_R)$ *such that the integral* $\int |\Psi_{config}(q_{1,config}, \ldots, q_{R,config})|^2) \, dq_{1,config}, \ldots, dq_{R,config}$, *converges."* ... . *"In the language of mathematics, the function space defined above is a Hilbert space. ..."*.

As a result of its abstract nature, configuration space (and/or Hilbert space) is sufficiently general to include additional degrees of freedom that are not associated with translational degrees of freedom. Such additional degrees of freedom may include the spin degrees of freedom of say the j-th particle. Furthermore, the generality of this abstract space allows Schrödinger's equation to be formulated to describe the dynamics of the spin-degrees of freedom, of say a 2-spin system, without reference to any spatial variables at all. This latter feature is an important asset for calculating the quantum–mechanical predictions for a Bell's theorem CH inequality test.

Despite its great generality, however, configuration space and the associated wave function have important limitations that of are worthy of note:

(1)   A position within the lab where the wave function is to be evaluated is not a degree of freedom of the system. Indeed, Ψ_{config} has no specified position within the lab where it is to be evaluated. It has the same value everywhere. Its value depends only on the system's "configuration", i.e. on the system's degrees of freedom.

(2)   The various theorems of vector analysis (e.g. Green's theorem(s), Gauss's theorem, Stokes theorem, etc.) do not apply to functions defined in configuration space, since these functions have no lab-space dependence.

(3)   There are no discernable waves that propagate in configuration space, like the waves that propagate in lab space. Dicke and Wittke (1960), and Merzbacher (1961,1970) both express alarm about this feature of configuration space (see the section "Born's Ambiguity's Misuse by the Lab-Space Formulation School"). To allow itself to be visualized, an entity like a wave must move in lab space. This fact becomes annoyingly apparent when two particles are needed for describing the helium atom.

(4)   Messiah (1961, p. 119) notes that *"The particle associated with the wave generally possesses neither a precise position nor a precise momentum."* Despite this claim, the single particle configuration-space wave function, $\Psi_{config,1}(\mathbf{r}_{1,config}; t)$, is defined in terms of the particle's *"precise position"* and thus depends on it's *"precise position."* This latter difficulty/ambiguity is somehow excused via the use of transformation theory.

(5)   Bjorken and Drell (1964, p. 2) state *"The wave function has no direct physical interpretation, however,* $|\Psi_{config}(q_{1,config}, ..., q_{n,config}, s_{1,config}, ..., s_{n,config}; t)|^2 \geq 0$ *is interpreted as the probability of the system having values of* $q_{1,config}, ...,$ $s_{n,config}$; *at time t. Evidently, this probability interpretation requires that the sum of positive contributions to* $|\Psi_{config}|^2$ *for all values of* $(q_{1,config}, ..., s_{n,config})$ *at time t to be finite for physically acceptable wave functions* Ψ_{config}*."* Bjorken and Drell, however, never explain the difference between a wave function that *"has no direct physical interpretation"* and a *"physically acceptable wave function".*

The section "Born's Argument-Space Ambiguity" further discusses some of these limitations and their unanticipated effects.

Schrödinger's Equation in Configuration Space

von Neumann (1932, 1955, Chap. 1) formulates Schrödinger's equation in configuration space. His formulation is similar to that by Messiah (1961, p. 71). It is derived using the Hamilton–Jacobi theory framework. Schrödinger's equation for a system with k configuration-space degrees of freedom, $q_{1,config}, ..., q_{k,config}$ is given first. The configuration space wave function for this system is then $\Psi_{config}(q_{1,config}, ..., q_{k,config};$ t), where $q_{1,config}, ..., q_{k,config}$ are dummy variables representing the k-degrees of freedom in the function. Schrödinger's equation for this system is

$$H_{config}\left(q_{1,config}, ..., q_{k,config}, -i\hbar\partial/\partial q_{1,config}, ..., -i\hbar\,\partial/\partial q_{k,config}\right)$$
$$\Psi_{config}\left(q_{1,config}, ..., q_{k,config}; t\right) = i\hbar\,\partial/\partial t\,\Psi_{config}\left(q_{1,config}, ..., q_{k,config}; t\right),$$

$$(3.22)$$

where H_{config} is the Hamiltonian operator for the system. In the particular case when the system consists of N mobile particles whose positions are specified by the $k = 3N$ coordinates, $q_{1,config}$, ..., $q_{3N,config}$, von Neumann sets $q_{3j-2,config}$, $q_{3j-1,config}$, $q_{3j,config}$ to be dummy variables representing the Cartesian coordinates of the j-th particle (with $j = 1, ..., N$), as per

$$q_{3j-2,config} \equiv x_{j,config}, \; q_{3j-1,config} \equiv y_{j,config}, \; \text{and} \qquad (3.23)$$
$$q_{3j,config} \equiv z_{j,config}.$$

Using these variables, one may formally define associated configuration space dummy variable vectors each specifying the position of the j-th particle within the lab as

$$\mathbf{r}_{j,config} \equiv \hat{\mathbf{e}}_x x_{j,config} + \hat{\mathbf{e}}_y y_{j,config} + \hat{\mathbf{e}}_z z_{j,config}, \qquad (3.24)$$

as well as with the momentum operator for the j-th particle,

$$\mathbf{p}_{j,config} \equiv -i\hbar \nabla_{j,config}. \qquad (3.25)$$

Here, the configuration-space gradient operator is defined by

$$\nabla_{j,config} \equiv \hat{\mathbf{e}}_x x_{j,config} \partial/\partial x_{j,config} + \hat{\mathbf{e}}_y \partial/\partial y_{j,config} + \hat{\mathbf{e}}_z \partial/\partial z_{j,config}. \qquad (3.26)$$

It should be noted that this operator is not at all the same as the lab-space gradient operator, defined above by equation (3.3).

The N-particle wave function for a system of N-particles is denoted by $\Psi_{config,N}(\mathbf{r}_{1,config}, \mathbf{r}_{2,config}, ..., \mathbf{r}_{N,config}, t)$. Note that the values of Ψ_{config} and $\Psi_{config,N}$ depend only upon the values of the various k degrees of freedom, and/or upon the various positions of the N particles in configuration space. Also note that the values of these functions have no explicit dependence on $\mathbf{r}_{Lab}$. That is to say, Ψ_{config} and $\Psi_{config,N}$ have the same value everywhere in lab space, and their values (also sometimes called amplitudes) are spatially constant. While the configuration-space gradient operator defined by (3.26), may have a formal symbolic appearance to the lab-space gradient operator defined above by (3.3), it is not the same operator! The configuration-space gradient operator produces the rate of change of its operand $\Psi_{config,N}$ with respect to a change in the position of particle j, while the lab space gradient operator produces the rate of change of its operand, Ψ_{Lab}, with respect to the lab-space position where said operand is to be evaluated.

It is emphasized that when the system consists of N mobile particles, the wave function, $\Psi_{config,N}(\mathbf{r}_{1,config}, \mathbf{r}_{2,config}, ..., \mathbf{r}_{N,config})$, has no evident spatial dependence, and no specified position in lab space where it is to be evaluated. That is, it does not depend on $\mathbf{r}_{lab}$. Instead, its various arguments indicate its dependence upon the positions, $\mathbf{r}_{1,config}$, $\mathbf{r}_{2,config}$, ..., $\mathbf{r}_{N,config}$, of the N particles. It is also important to note that no special status is given to particle 1 (of N) and to the associated first argument of $\Psi_{\text{config},N}$, even for the special case $N = 1$. Thus, even for a single particle system, the first argument, $\mathbf{r}_{1,\text{config}}$, still specifies the position of the one and only one particle, and not the position within in the lab where the function, $\Psi_{\text{config},1}$, is to be evaluated.

As a result, for a system with k degrees of freedom,

$$\nabla_{\text{Lab}} \Psi_{\text{config}} \left(q_{1,\text{config}}, \ldots, q_{k,\text{config}}; t \right) = 0 \qquad (3.27)$$

holds, and for a system of N particles

$$\nabla_{\text{Lab}} \Psi_{\text{config},N} \left(\mathbf{r}_{1,\text{config}}, \mathbf{r}_{2,\text{config}}, \ldots, \mathbf{r}_{N,\text{config}}, t \right) = 0 \qquad (3.28)$$

holds for any N, including for the special case, $N = 1$.

By using an operator substitution trick similar to that used by the lab space formulation, the Hamiltonian operator for an N-particle (spinless) system is given by

$$H_{\text{config},N}(\mathbf{r}_{1,\text{config}}, \mathbf{r}_{2,\text{config}}, \ldots, \mathbf{r}_{N,\text{config}}, -i\hbar\nabla_{1,\text{config}}, -i\hbar\nabla_{2,\text{config}}, \ldots,$$

$$-i\hbar\nabla_{N,\text{config}}) \Psi_{\text{config},N} \left(\mathbf{r}_{1,\text{config}}, \mathbf{r}_{2,\text{config}}, \ldots, \mathbf{r}_{N,\text{config}}; t \right)$$

$$\equiv \sum_{j=1,N} [- \left(\hbar^2/2m \right) \nabla^2_{j,\text{config}} + V(\mathbf{r}_{j,\text{config}})] + \sum_{j,k=1,N} V_{j,k} \left(\mathbf{r}_{j,\text{config}}, \mathbf{r}_{k,\text{config}} \right) .$$

$$(3.29)$$

The last term in (3.29) allows for interactions between the particles. Schrödinger's time-dependent equation for an N-particle system takes the form

$$H_{\text{config},N}(\mathbf{r}_{1,\text{config}}, \mathbf{r}_{2,\text{config}}, \ldots, \mathbf{r}_{N,\text{config}}, -i\hbar\nabla_{1,\text{config}}, -i\hbar\nabla_{2,\text{config}}, \ldots,$$

$$-i\hbar\nabla_{N,\text{config}}) \Psi_{\text{config},N} \left(\mathbf{r}_{1,\text{config}}, \mathbf{r}_{2,\text{config}}, \ldots, \mathbf{r}_{N,\text{config}}; t \right) = \qquad (3.30)$$

$$i\hbar\partial/\partial t \, \Psi_{\text{config},N} \left(\mathbf{r}_{1,\text{config}}, \mathbf{r}_{2,\text{config}}, \ldots, \mathbf{r}_{N,\text{config}}; t \right) .$$

In the $N = 1$ special case, the single particle wave Schrödinger's time-dependent equation is

$$H_{\text{config},N=1}(\mathbf{r}_{1,\text{config}}, -i\hbar\nabla_{1,\text{config}}) \Psi_{\text{config},1} \left(\mathbf{r}_{1,\text{config}}; t \right) = i\hbar\partial/\partial t \, \Psi_{\text{config},1} \left(\mathbf{r}_{1,\text{config}}; t \right) .$$

$$(3.31)$$

Finally, stationary state solutions, $\Psi_{\text{config},N}(\mathbf{r}_{1,\text{config}}, \mathbf{r}_{2,\text{config}}, \ldots, \mathbf{r}_{N,\text{config}})$, are readily found by factoring out the time dependence, as per

$$\Psi_{\text{config},N} \left(\mathbf{r}_{1,\text{config}}, \mathbf{r}_{2,\text{config}}, \ldots, \mathbf{r}_{N,\text{config}}, t \right)$$

$$= \Psi_{\text{config},N} \left(\mathbf{r}_{1,\text{config}}, \mathbf{r}_{2,\text{config}}, \ldots, \mathbf{r}_{N,\text{config}} \right) \exp \left(-iEt/\hbar \right) , \qquad (3.32)$$

whereupon Schrödinger's N-particle and single particle time-independent equations in configuration space are given by

$$
H_{\text{config,N}}(\mathbf{r}_{1,\text{config}}, \mathbf{r}_{2,\text{config}}, \ldots, \mathbf{r}_{N,\text{config}}, -i\hbar\nabla_{1,\text{config}}, -i\hbar\nabla_{2,\text{config}}, \ldots,
$$
$$
-i\hbar\nabla_{N,\text{config}}) \Psi_{\text{config,N}}(\mathbf{r}_{1,\text{config}}, \mathbf{r}_{2,\text{config}}, \ldots, \mathbf{r}_{N,\text{config}}) \tag{3.33}
$$
$$
= E_{\text{tot-N}}\Psi_{\text{config,N}}(\mathbf{r}_{1,\text{config}}, \mathbf{r}_{2,\text{config}}, \ldots, \mathbf{r}_{N,\text{config}}),
$$

and

$$
H_{\text{config,N=1}}(\mathbf{r}_{1,\text{config}}, -i\hbar\nabla_{1,\text{config}})\Psi_{\text{config,1}}(\mathbf{r}_{1,\text{config}}) = E_{\text{tot-1}}\Psi_{\text{config,1}}(\mathbf{r}_{1,\text{config}}; t).
$$
$$
\tag{3.34}
$$

The quantities $E_{\text{tot-1}}$ and $E_{\text{tot-N}}$ are energy eigenvalues for single particle and N particle systems, respectfully.

Calculation of Probabilities Using the Configuration-Space Formulation

Messiah (1961, pp. 126–127) shows how to calculate probabilities within the configuration-space formulation using Born's statistical interpretation. The joint probability density for finding the N particles at a specified set of N positions within the lab within in the differential volume elements, $d^3\mathbf{r}_{1,\text{config}} \equiv dx_{j,\text{config}} dy_{j,\text{config}} dz_{j,\text{config}}$, etc. $d^3\mathbf{r}_{2,\text{config}}$, is found by evaluating the absolute square of the configuration-space wave function at those N positions within the lab. Thus the probability density, $P_{1,2,\ldots,N}$, for finding a particle 1 at a detector that is positioned at $\mathbf{r}_{1,\text{det}}$, and also finding particle 2 at a second detector that is positioned at $\mathbf{r}_{2,\text{det}}$, etc., between times t_f and $t_f + dt$, is found by setting the dummy variable arguments in $\Psi_{\text{config,N}}$, to $\mathbf{r}_{1,\text{config}} = \mathbf{r}_{1,\text{det}}$, $\mathbf{r}_{2,\text{config}} = \mathbf{r}_{2,\text{det}}$, etc. and then by calculating it via

$$
P_{1,2,\ldots,N}(\mathbf{r}_{1,\text{config}} = \mathbf{r}_{1,\text{det}}, \mathbf{r}_{2,\text{config}} = \mathbf{r}_{2,\text{det}}, \ldots, \mathbf{r}_{N,\text{config}} = \mathbf{r}_{N,\text{det}}; t = t_f)
$$
$$
= \left|\Psi_{\text{config,N}}(\mathbf{r}_{1,\text{config}} = \mathbf{r}_{1,\text{det}}, \mathbf{r}_{2,\text{config}} = \mathbf{r}_{2,\text{det}}, \ldots, \mathbf{r}_{N,\text{config}} = \mathbf{r}_{N,\text{det}}; t = t_f)\right|^2.
$$
$$
\tag{3.35}
$$

Normalization of this probability density is then

$$
\int d^3\mathbf{r}_{1,\text{config}}., d^3\mathbf{r}_{2,\text{config}}, \ldots, d^3\mathbf{r}_{N,\text{config}}P_{1,\ldots,N}(\mathbf{r}_{1,\text{config}}, \mathbf{r}_{2,\text{config}}, \ldots, \mathbf{r}_{N,\text{config}}, t)
$$
$$
= \int d^3\mathbf{r}_{1,\text{config}}, d^3\mathbf{r}_{2,\text{config}}, \ldots, d^3\mathbf{r}_{N,\text{config}}\left|\Psi_{\text{config,N}}(\mathbf{r}_{1,\text{config}}, \mathbf{r}_{2,\text{config}}, \ldots, \mathbf{r}_{N,\text{config}})\right|^2, t)
$$
$$
= 1.
$$
$$
\tag{3.36}
$$

For say an $N = 2$ particle system, the joint probability density for finding the particles at the positions $r_{1,\text{config}} = r_{1,\text{det}}$ and $r_{2,\text{config}} = r_{2,\text{det}}$ is given by

$$
\begin{aligned}
&P_{1,2}\left(r_{1,\text{config}} = r_{1,\text{det}}, r_{2,\text{config}} = r_{2,\text{det}}; \; t = t_f\right) \\
&= \left|\Psi_{\text{config},2}\left(r_{1,\text{config}} = r_{1,\text{det}}, r_{2,\text{config}} = r_{2,\text{det}}; \; t = t_f\right)\right|^2.
\end{aligned}
\tag{3.37}
$$

The individual probability density for finding particle 1 at position $r_{1,\text{config}} = r_{1,\text{det}}$, without caring about the position particle 2 is given by integrating over the possible positions, $r_{2,\text{config}}$, for particle 2,

$$
\begin{aligned}
P_1\left(r_{1,\text{config}} = r_{1,\text{det}}; \; t = t_f\right) &= \int P_{1,2}\left(r_{1,\text{config}} = r_{1,\text{det}}, r_{2,\text{config}}; \; t = t_f\right) d^3 r_{2,\text{config}} \\
&= \int \left|\Psi_{\text{config}, N=2}\left(r_{1,\text{config}} = r_{1,\text{det}}, r_{2,\text{config}}; \; t = t_f\right)\right|^2 d^3 r_{2,\text{config}},
\end{aligned}
\tag{3.38}
$$

and the corresponding individual probability density for finding particle 2 at position $r_{2,\text{config},2}$ is similarly given by

$$
\begin{aligned}
P_2\left(r_{2,\text{config}} = r_{2,\text{det}}; \; t = t_f\right) &= \int P_{1,2}\left(r_{1,\text{config}}, r_{2,\text{config}} = r_{2,\text{det}}; \; t = t_f\right) d^3 r_{1,\text{config}} \\
&= \int \left|\Psi_{\text{config}, N=2}\left(r_{1,\text{config}}, r_{2,\text{config}} = r_{2,\text{det}}; \; t = t_f\right)\right|^2 d^3 r_{1,\text{config}}.
\end{aligned}
\tag{3.39}
$$

Factorization of Schrödinger's N-particle Configuration-Space Wave Function

An important "factorization criterion" for Schrödinger's N-particle configuration-space wave function is given in many textbooks to argue that the N-particle configuration-space wave function, $\Psi_{\text{config},N}(r_{1,\text{config}}, r_{2,\text{config}}, \ldots, r_{N,\text{config}}, t)$ is a proper generalization of the single particle wave function, formulated in either lab or configuration space. Unfortunately, it really only works for the latter formulation, and even then, it only works in the unrealistic special situation where any and all interaction between these particles exactly vanishes. It is readily demonstrated to apply to both the time-independent and time-dependent configuration-space Schrödinger's equations. The demonstration is usually given for two-particle systems, but it may be generalized to apply to N-particle systems. It is also typically given for systems for which the degrees of freedom are position and momentum, as is done here. However, it also may be generalized to apply to systems using other degrees of freedom, such as particle spin. The argument is given, for example, by Dicke and Wittke (1960, p 111) for single particle wave functions in lab space, and by Messiah (1961, pp. 127–128), for single particle wave functions in configuration space. Messiah's treatment is followed here.

Consider a two-particle system for which the following conditions hold:

(1)   The interaction between the two particles vanishes. That is, the interaction potential between them is assumed to be time independent and vanishes for all time, as per

$$V_{1,2}(\mathbf{r}_{1,\text{config}}, \mathbf{r}_{2,\text{config}}) = 0, \tag{3.40}$$

(2)   The two-particle configuration-space Hamiltonian has the form of a sum of two terms.

$$\begin{aligned} H_{\text{config},N=2}(\mathbf{r}_{1,\text{config}}, \mathbf{r}_{2,\text{config}}, -i\hbar\nabla_{1,\text{config}}, -i\hbar\nabla_{2,\text{config}}) \\ = H_{\text{particle1},\text{config},N=1}(\mathbf{r}_{1,\text{config}}, -i\hbar\nabla_{1,\text{config}}) \\ + H_{\text{particle2, config},N=1}(\mathbf{r}_{2,\text{config}}, -i\hbar\nabla_{2,\text{config}}), \end{aligned} \tag{3.41}$$

where $H_{\text{particle1},\text{config},N=1}$ and $H_{\text{particle2, config},N=1}$ are the Hamiltonians for the two individual particles.

(3) $\Psi_{\text{config},1}(\mathbf{r}_{1,\text{config}})$ and $\Psi_{\text{config},2}(\mathbf{r}_{2,\text{config}})$ are energy eigenfunctions of the single particle Schrödinger's equations, (3.34) for particles 1 and 2, respectively with respective (single particle) energy eigenvalues E_1, and E_2.

(4) $\Psi_{\text{config},N=2}(\mathbf{r}_{1,\text{config}}, \mathbf{r}_{2,\text{config}})$ is an energy eigenfunction of the $N = 2$ particle Schrödinger's equation, (3.33) with the energy eigenvalue $E_{\text{tot-}N=2} = E_1 + E_2$.

When conditions (1)–(4) hold, it is readily shown that the product of the two single-particle eigenfunctions is then an eigenfunction of the $N = 2$ particle Schrödinger's equation, as per

$$\Psi_{\text{config},1}(\mathbf{r}_{1,\text{config}})\,\Psi_{\text{config},2}(\mathbf{r}_{2,\text{config}}) = \Psi_{\text{config},N=2}(\mathbf{r}_{1,\text{config}}, \mathbf{r}_{2,\text{config}}). \tag{3.42}$$

Messiah (1961, p. 128) notes that this "factorization criterion", (3.42), persists in time, so that, if initially at $t = t_0$, the time-dependent two-particle wave function can be factored, then this condition persists for $t > t_0$, thus

$$\begin{aligned} \Psi_{\text{config},N=2}(\mathbf{r}_{1,\text{config}}, \mathbf{r}_{2,\text{config}}, t) = \Psi_{\text{particle1},\text{config, }N=1}(\mathbf{r}_{1,\text{config}}, t) \\ \Psi_{\text{particle2},\text{config, }N=1}(\mathbf{r}_{2,\text{config}}, t), \end{aligned} \tag{3.43}$$

holds for $t > t_0$. Messiah (1961) also notes that when the wave function thusly factors, then the joint probability density, as defined above by (3.37), also factors,

$$P_{1,2}(\mathbf{r}_{1,\text{config}}, \mathbf{r}_{2,\text{config}}) = P_1(\mathbf{r}_{1,\text{config},2})\,P_2(\mathbf{r}_{2,\text{config},2}), \tag{3.44}$$

and the particles are statistically independent. Messiah's "factorization criterion", (3.43), is equivalent to particle independence.

Probabilities of separated measurements specified by (3.43), (3.38), and (3.39) may be compared with similar probabilities calculated via Local Realism using (3.4), (3.5) and (3.7). Note that the particle-independence factorization condition, (3.44), is similar to (3.6), used by Local Realism above in the section "Bell–Clauser–Horne–Shimony Local Realism". There is no similar joint probability of separated measurements for a lab-space formulation because the lab space formulation describes only single-particle systems.

Wave-Function Factorization or Not!

Lab-space formulations of quantum mechanics (e.g. Dicke and Wittke (1960, p 111)) fallaciously use the factorization argument of the preceding section to argue that $\Psi_{\text{config},N=2}(\mathbf{r}_{1,\text{config}}, \mathbf{r}_{2,\text{config}})$ is a proper generalization of single-particle lab-space wave function, $\Psi_{\text{lab}}(\mathbf{r}_{\text{lab}})$, so as to allow the lab space formulation to handle N-particle systems. The argument immediately crumbles, however, when one discovers (as Dicke and Wittke note) that by doing so, the wave function's wave-like properties are then lost. Dicke and Wittke, however fail to note that not only is the argument fallacious, it is also incomplete and ignores other even worse problems. A more complete argument shows that, in the fully general case, even when there is no interaction between the particles, then (3.42), (3.43), and (3.44) do not necessarily hold. Similarly, Messiah does not comment that, even if the particles do not interact after $t = t_0$, but if they have ever interacted in the past, even only slightly prior to $t = t_0$, then the N-particle wave function does not factor, and the particles are not statistically independent. An arbitrary choice with t_0 finite, and/or with V_{12} possibly time dependent then belies the possibility that the particles may have been interacting at some time since the beginning of the universe.

More importantly, the factored forms (3.42) and (3.43) are not the only solutions to Schrödinger's equation(s), even when V_{12} is always time independent. Important other solutions exist that apply to entangled particle states, even when the particles are non-interacting. An important property of entangled state solutions then provides the converse of Messiah's independent particle property. That is, entanglement persists in time, even when there is no longer any interaction between the particles. Messiah's argument should correspondingly be modified to read: *Once independent and with no interaction, then always independent. Conversely, once entangled and even with no interaction, then always entangled.*

The usage of the above wave-function factorization argument by authors to bolster their claims that the configuration-space wave function, $\Psi_{\text{config},N}(\mathbf{r}_{1,\text{config}}, \mathbf{r}_{2,\text{config}}, ..., \mathbf{r}_{N,\text{config}}, t)$, is an appropriate N-particle generalization of the single-particle lab-space wave function, $\Psi_{\text{Lab}}(\mathbf{r}_{\text{Lab}}, t)$, thus totally looses its credibility when the existence of these other entangled state solutions is revealed. If Messiah's argument had been completed correctly, then the appropriateness of $\Psi_{\text{config},N}(\mathbf{r}_{1,\text{config}}, \mathbf{r}_{2,\text{config}}, ...,$

$\mathbf{r}_{N,config}$, t) as a generalization of the $N = 1$ single particle wave function would be immediately found wanting.

Curiously, despite entanglement's absence in the arguments of the above-mentioned authors, it has been around for a long time. For example, Condon and Shortley (1964) formulate Schrödinger's equation exclusively using configuration space. Their application of quantum mechanics is to many-particle systems. They discuss the factorization property, along with entanglement, and provide a practical use for it. Typically, independent single-electron Hamiltonians and factored wave functions are used there to provide an approximate first-guess solution to an N-particle problem. The interaction potential is then used as a small perturbation to provide an improved solution.

Entangled particle states typically occur when there are degenerate solutions to the single-particle Schrödinger's equations, (3.34), as may occur when there are additional degrees of freedom such as spin for the particles. Entangled particles do not need to be so-called "identical particles". The versatility of the configuration space formulation now becomes particularly useful. Suppose that the single particle wave functions, $\Psi_{particle1,config, N=1}(\mathbf{r}_{1,config}, s_1)$ and $\Psi_{particle2,config,N=1}(\mathbf{r}_{2,config}, s_2)$, include additional degrees of freedom, s_1 and s_2. Suppose also that these degrees of freedom refer to discrete-state variables, such as spin. This degree of freedom for spin ½ has only two allowed states. If allowed integer dummy index values, $s_1, s_2. = 1, 2$ are used, then both associated indexed values may be displayed together as a two component column vector.[19] Alternatively, the symbols, s_1, s_2, are sometimes used as dummy indicies. They may each take on one of the two values, $\uparrow$ or $\downarrow$. (Born uses the values $+$ and $-$.) Finally, assume that the energy eigenstates are degenerate and both particles have the same energy, irrespective of the value of s_j. It is then straightforward to show that

$$\Psi_{config,N=2-particles}(\mathbf{r}_{1,config}, \mathbf{r}_{2,config}) = \Sigma_{s1=\uparrow,\downarrow} \Sigma_{s2=\uparrow,\downarrow} a_{s1,s2}$$
$$\Psi_{particle1,config, N=1}(\mathbf{r}_{1,config}, s_1) \Psi_{particle2,config, N=1}(\mathbf{r}_{2,config}, s_2). \tag{3.45}$$

The form (3.45) is no longer a simple product, as per (3.42), but is instead now a sum of products of eigenfunctions, with associated amplitudes $a_{s1, s2}$. In some special cases (and only in these cases), such as when all of the various $a_{s1, s2}$ vanish except one, we have a simple product state like (3.42).

The form (3.45) is useful for demonstrating the Einstein–Podolsky–Rosen paradox via Bohm's two entangled-spins *Gedankenexperiment*. (See Clauser (2017).) There, each of the particles may proceed along one of two different paths through an associated Stern-Gerlach apparatus. When, for example, the spatial dependence

---

[19] This alternative form is used by Born (1933, 1935, 1969, p.188). Referring to spin, he comments *"We can take this new degree of freedom into account formally, by introducing besides the ordinary co-ordinates an additional co-ordinate σ, which can take only two values together; ... We thus obtain a wave function which now depends on five co-ordinates: $\Psi = \Psi(x,y,z,t,\sigma)$. It suggests itself, however, to split up this wave function into two components..."* Nowhere, however, does Born admit that this set of *"five-coordinates"* is, in fact, in configuration space. (See the section "Born's Argument-Space Ambiguity".).

in (3.45) is not needed, and only the spin dependence is needed or relevant, then $\Psi_{\text{particle1,config, N=1}}(s_1)$ and $\Psi_{\text{particle2,config,N=1}}(s_2)$ can be used to represent the single particles' wave functions. The two-particle entangled-state wave function can then be represented as

$$\Psi_{\text{config,N=2-particles}} = \Sigma_{s1=\uparrow,\downarrow} \Sigma_{s2=\uparrow,\downarrow} a_{s1,s2}$$
$$\Psi_{\text{particle1,config, N=1}}(s_1) \; \Psi_{\text{particle2,config, N=1}}(s_2). \tag{3.46}$$

It should be noted in passing that equations (3.27) and (3.28) hold for both forms (3.45) and (3.46), so that no conserved probability current can be calculated for either of these two particle systems.

It also should also be noted in passing that Messiah does not give any criterion regarding how small $V_{1,2}$ must be for it to be considered "negligible", eventhough his argument requires that it vanish completely, as per condition (1), above. Unfortunately, particle independence will decay exponentially in time, whereby Messiah's persistence in time of particle independence is then actually really only transitory, even for very small $V_{1,2}$. Furthermore, Spitzer (1956) points out that the cross-section for classical Coulomb scattering between two charged particles diverges and becomes infinite at small scattering angles. Thus, all charged particles always continuously interact with each other, by at least a very small amount. Moreover, all charged particles have been thusly interacting with each other since the beginning of the universe. Additionally, a significant fraction of the universe is comprised of charged particles (protons and electrons), so it would seem short-sighted to ignore their interaction completely. Correspondingly, one may conclude, quantum mechanically, that all charged particles are always entangled, at least by a very small amount, especially since $t = t_0$ may be taken arbitrarily to be a time in the very distant past. Given that for any non-vanishing $V_{1,2}$, particle independence decays in time, then entanglement correspondingly grows in time. Eventually, entanglement always wins out over independence. One may correspondingly wonder, "*Is being a little bit entangled like being a little bit pregnant? Perhaps for both cases, the importance of both conditions depends upon how long you wait.*"

Born's Argument-Space Ambiguity

Unfortunately, there are important discrepancies between the lab-space and configuration-space formulations of quantum mechanics. These discrepancies can be traced to originate with a somewhat hidden ambiguity that was introduced by Born (1933, 1935, 1969) in his seminal book. While promoting his "statistical interpretation", and discussing the configuration-space description of two-particle Rutherford scattering, Born (1933, 1935, 1969, pp. 95–96) states

> "*There are grounds for the conviction of the correctness of the principle of associating wave amplitude with number of particles (or probability). In this picture, the particles are regarded*

as independent of each other. If we take their mutual action into account, the pictorial view is to some extent lost again. We have then two possibilities. Either we use waves in spaces of more than three dimensions (with two interacting particles we would have 2 X 3 = 6 coordinates), or we remain in three-dimensional space but give up the simple picture of the wave amplitude as an ordinary physical magnitude and replace it with a purely abstract mathematical concept, (the second quantization of Dirac, Jordan) into which we cannot enter. ... This is for us the really important question, for clearly enough the corpuscular and wave ideas cannot be fitted together into a homogeneous theoretical formalism without giving up some fundamental principles of the classical theory. The unifying concept is that of probability."

Born thus takes $\Psi_{lab}(\mathbf{r}_{lab}, t)$ to be an *"ordinary physical magnitude"* in lab space, i.e. in what he calls *"three-dimensional space"*. He also claims that we can *"remain in three-dimensional space"* i.e. use the amplitude $\Psi_{lab}(\mathbf{r}_{lab}, t)$ in lab space as long as we use his probability "concept". He thereby claims that Ψ_{lab} and $\Psi_{config,N=2}$, are equivalent, and that both have an *"ordinary physical magnitude"*, even though they are differently defined for the two different argument spaces. He asserts that this is true simply because they are both defined to yield the same probability. Added together, he is saying that $\Psi_{lab}(\mathbf{r}_{lab}, t)$ and $\Psi_{config,N=2}(\mathbf{r}_{1,config,2}, \mathbf{r}_{2,config,2}, t)$ and their associated argument spaces, $\mathbf{r}_{lab}$ and $\mathbf{r}_{1,config,2}$, $\mathbf{r}_{2,config,2}$, are equivalent and interchangeable, He doesn't seem to notice that there are, in fact, two (or more) different *"three dimensional spaces"* to choose from—the argument spaces for $\mathbf{r}_{lab}$ and for $\mathbf{r}_{1,config,1}$— and he correspondingly does not distinguish between them.

Herein he also creates an ambiguity regarding the meanings and uses of the arguments, $\mathbf{r}_{lab}$ and $\mathbf{r}_{1,config,2}$, $\mathbf{r}_{2,config,2}$, and of the associated wave function's, Ψ_{lab} and $\Psi_{config,N=2}$. He rejects the idea that either wave function's meaning is that used in quantum field theory (as discussed below in the sections "Quantum Field Theory 1—Quantization of Known Classical Fields" and "Quantum Field Theory 2—Second Quantization of Wave-Functions"), because it is *"a purely abstract mathematical concept"*. Motivated by his evident distaste for purely abstract mathematical concepts, and given his proclaimed equivalence of the spaces, Born appears to say that the use of lab space may be chosen simply as a matter of taste. By doing so, he misses the fact that configuration space and the associated wave function, $\Psi_{config,N=2}$, are also themselves purely abstract mathematical concepts, which he admits that he dislikes.

Born's quote creates an ambiguity. It has thus pronounced the lab-space and configuration space formulations to be equivalent and interchangeable. Taken at face value, any such claim is clearly unfounded. The spaces are indeed very different, as has been demonstrated above. Born and his lab-space-formulation followers similarly appear to have missed (or chosen to ignore) the fact that configuration-space is not as versatile as they may have assumed. It cannot do two incompatible things simultaneously. The wave function's argument clearly does not encompass the ability to represent both the position in lab space where Ψ is to be evaluated, and, at the same time, also simultaneously to represent the configuration of the system being described. <u>Simply put, it cannot be multiply defined.</u> The two formulations and their associated argument spaces are clearly not equivalent. The two wave functions are also not equivalent. In addition, they are not equal. As noted above in the section

"The Configuration-Space Formulation of Quantum Mechanics", there is no spatial dependence for Ψ_{config}. Furthermore, recall that there is no special meaning for the various arguments, especially for the first argument of $\Psi_{config,N=1}$, in the special case $N = 1$. It is not the same argument as that of Ψ_{lab}. As a result, we have in general, $\mathbf{r}_{lab} \neq \mathbf{r}_{config,1,1}$, even though they are both formally three-dimensional.

Correspondingly, for any number of particles, N, described by $\Psi_{config,N}$, we have (as per Eqs. (3.27) and (3.28))

$$\nabla_{Lab}\Psi_{config,N} = 0, \text{ and } \nabla_{Lab}\Psi_{config,1} = 0. \tag{3.47}$$

Gauss's theorem, Green's theorem and the conservation law (3.21) correspondingly do not apply to $\Psi_{config,N}$, or to $\Psi_{config,1}$. Since the arguments and definitions of the wave functions Ψ_{lab} and $\Psi_{config,N=2}$, and the definitions of their associated arguments, $\mathbf{r}_{lab}$ and $\mathbf{r}_{1,config,2}$, $\mathbf{r}_{2,config,2}$ are all very different, one also must conclude further that

$$\Psi_{lab} \neq \Psi_{config,N=2}, \text{ and } \mathbf{r}_{lab} \neq \mathbf{r}_{j,config,N}, \tag{3.48}$$

hold for any j and/or N, including the special cases $j = N = 1$.

As we have noted in the section "Introduction - What quantum mechanics is Not", Born's ambiguity is produced by a sneaky slight-of-hand. The ambiguity manifests itself by using the same ambiguous (multiply defined) symbol $\mathbf{r}$ (or symbols $\mathbf{r}_1$, $\mathbf{r}_2$, …) to represent very different quantities (or sets of quantities) with different meanings altogether. Then, via this subterfuge two different equations (or sets of equations) that are formally the same in their appearance(s) are produced and claimed to govern nature. Both equations (or sets of equations) use the common ambiguous symbol (or symbols). Presto, the equations are thereby claimed to be equivalent, when in reality they are not. The misdirection and substitution are done so seamlessly that no one is aware of the prestidigitation that has passed. In particular, the configuration-space single-particle Schrödinger's equations (3.31) and (3.34) and the lab-space Schrödinger's equations (3.14) and (3.16) are not the same, despite their formal similarities. Indeed, they are formulated in very different argument spaces.

Born's Ambiguity's Misuse by the Lab-Space Formulation School

Unfortunately, a lab-space formulation of wave mechanics is found to suffer from several serious deficiencies. Paramount among these is that it can only describe single particle systems. We have just noted that there does not appear to be any rigorous (local) method to allow it to be extended to $N \geq 2$ particle systems. As a result, a lab-space formulation cannot describe entanglement, as is needed for describing the spectra of atoms with more than one electron and for Bell inequality tests. The inability of the lab-space formulation (outlined above in the section "The lab-space formulation of quantum mechanics") to find a lab-space description of two-particle systems becomes immediately apparent when one is faced with treating the helium atom.

Lab space advocates (see the section "The Lab-Space Formulation of Quantum Mechanics") formulate Schrödinger's equation, specifically to allow wave propagation in lab space via (3.9). Correspondingly, a lab-space wave function clearly <u>can</u> demonstrate actual wave motion, as is depicted in Figs. 3.1–3.7, for matter waves. Correspondingly, said wave function can be used as a model for experimental observations, as are described, for example, by French and Taylor (1958, 1978). Unfortunately, a configuration-space wave function, like $\Psi_{\text{config},N=2}$ <u>cannot</u> demonstrate wave motion, as was noted by Dicke and Wittke (1960) and by Merzbacher (1961, 1970). The whole lab-space model depicted in Figs. 3.1–3.7 now begins to crumble.

While attempting to describe the helium atom, Merzbacher (1961, 1970, p. 347) notices the argument-space discrepancy between single particle and two particle wave functions. He comments *"Since Ψ is now a function of two different points in space, it can no longer be pictured as a wave in the naïve sense which we found so useful in the early chapters of this book. Instead, Ψ for two particles must sometimes be considered a wave in a six dimensional configuration space of the coordinates* $\mathbf{r}_{1,\text{config},2}$ *and* $\mathbf{r}_{2,\text{config},2}$.*"* Merzbacher, however, does not offer any suggestions for restoring the lost utility of his now discredited early chapters. He also says that viewing a wave function as a description of waves propagating in lab space is naïve. He does not, however, offer a more sophisticated viewpoint.

Similarly, Dicke and Wittke (1960, p. 110) comment *"Hence the wave function has the form $\Psi = \Psi_{\text{config},N=2}(\mathbf{r}_{1,\text{config},2}, \mathbf{r}_{2,\text{config},2}, t)$. Note that this function can hardly be interpreted as a physical wave moving in ordinary three-dimensional space. It has the form of a wave moving in a six-dimensional space. Since this is the analog of the wave function for a one-particle system, it is clear that physical wavelike properties which a single particle wave function exhibits are properties which are to be ascribed to one-particle systems only. In other words, Ψ is a physical wave only to the extent that it can be associated with the motion of single particles."*

Almost all authors of lab-space-formulation based textbooks incorrectly enlist Born's argument-space ambiguity, along with Messiah's particle-independence criterion, as described above in the section "Factorization of Schrödinger's N-particle Configuration-Space Wave Function", to fill the N-particle description void left by the lab-space formulation. Correspondingly, it seems probable that the ambiguity introduced by Born is the cause of the conceptual teachings by the lab-space formulation school and the configuration-space school becoming bifurcated in subsequently authored textbooks. Accordingly, Dicke and Wittke (1960, pp. 110–111) and French and Taylor (1958, 1978, pp. 558–560), proclaim (following Born's quote) that the 2-particle configuration-space wave function, $\Psi_{\text{config},2}(\mathbf{r}_{1,\text{config}}, \mathbf{r}_{2,\text{config}}, t)$, is an appropriate 2-particle "generalization" of the single-particle lab-space wave function $\Psi_{\text{Lab}}(\mathbf{r}_{\text{Lab}}, t)$. They do so by simply swapping dummy index definitions, $\mathbf{r}_{1,\text{config}} \leftrightarrow \mathbf{r}_{1,\text{lab}}, \mathbf{r}_{2,\text{config}} \leftrightarrow \mathbf{r}_{2,\text{lab}}$, and ignoring the implications of said swap. Our use herein of "Lab" and "config" subscripts immediately reveals the error in this procedure. The argument spaces for lab-space and configuration-space wave functions are clearly not the same. The meanings of the arguments in the wave functions are also not the same. One wave function clearly cannot be considered to be the same as the other, simply because it has the same number of three- "dimensional" arguments, or because its absolute square yields a probability. The associated lab-space and configuration-space Schrödinger's equations, (3.14) and (3.31), while formally similar in appearance (without the distinguishing subscripts), are actually very different, especially in their meanings. Thus, there is no proper N-particle generalization of the lab-space Schrödinger's equation, (3.14). A lab-space formulation is thus limited to the treatment of single particle motion, as was observed and emphasized by Dicke and Wittke (1960).

Alas, the conceptual model that lab-space formulations promote and use is found to be untenable and unable to describe even a helium atom. One might be tempted to justify keeping the lab-space formulation of quantum mechanics, because it perhaps provides an approximation to a better theory. However, changing from lab space hardly can be considered an approximation of configuration space. Is an apple an approximation of an orange? Nonsense! That justification thus fails miserably. Unfortunately, the price one must pay is (sadly) that the configuration-space formulation requires a highly abstract mathematical formalism that is difficult to understand, and that provides no conceptual model to allow its inner workings to be visualized.

Born's Ambiguity's Misuse by the Configuration-Space School

Another important discrepancy between the two schools that is caused by Born's ambiguity occurs with the derivation and use of his conserved probability current. A requirement for any formulation of quantum mechanics is the conservation of particle number. Born had demonstrated this requirement to hold for Rutherford scattering using the lab space formulation. Configuration space advocates are correspondingly obliged to offer a similar assurance. They generally do so by using Born's ambiguity to demonstrate conservation of particle number, by using what Messiah (1961, pp. 119–122) calls a "*conserved current concept*". Messiah claims "*The property of conservation of the norm has a simple interpretation if one introduces the notion of current.*" His derivation follows exactly that by Born, and totally ignores the fact that Green's theorem does not apply in configuration space. He correspondingly ignores Eq. (3.47). Despite this oversight, he and other authors (see, for example, Landau and Lifshitz (1958, 1965)) claim "conservation of probability in configuration space", whatever that means. (Messiah does not tell us what it means.)

Messiah (1961, pp. 222–223) also provides a second, alternative derivation of the conservation equation for the current in configuration space using the classical Hamilton's function S. He notes that "*In the classical approximation, Ψ_{config} describes a fluid of non-interacting particles of mass m, (statistical mixture) and subject to the potential, $V(r_{config})$, the density and current density of this fluid at each* *point of space are at all times respectively equal to the probability density* P_{config} *and the probability current density* J_{config} *of the quantum particle at that point.*

On p. 224, he further notes that it is "... *valid for systems with any number of dimensions. The density* $P_{config} = |\Psi_{config}|^2$ *is a well defined function of configuration space; similarly the current* J_{config} *is a well defined vector field of that space.*"

What is meant by his concept of a "*vector field ... in a configuration space ... with any number of dimensions*" is never explained. Messiah also defines what he means by a "*vector field in a space with* "*any number of dimensions*", especially when some of those "dimensions" may correspond to non-classical degrees of freedom like spin. Whatever his definition might be, it is clearly very different from the definition of a vector field in lab-space, as is given above in the above section "Laboratory Space and Classical Fields". Messiah has definitely exploited Born's ambiguity to its limits. It should also be noted that Messiah admits (p. 121), "*Of course, the analogy between this probability fluid should not be pushed too far. All pictures based on this analogy contain no more than the property (IV.11).*" [Messiah's Eq. (IV.11), is the conservation law for a classical fluid flowing in lab space.]

In a fashion similar to that used by Born (i.e. use of the probability current's conservation to explain particle flux conservation in Rutherford scattering), Messiah (1961, pp. 369–380) also uses his current "concept", along with Born's ambiguity to describe particle scattering, and to define scattering cross-sections. His description of particle scattering, along with the associated illustrative Figures [Messiah (1961, pp. 374–375)], however, describes waves propagating in lab space, and particles moving in lab space, similarly to those in Figs. 3.1–3.7, above.

Born's Conserved Probability Current as Re-interpreted Using the Configuration-Space Formulation

As noted above, various textbooks firmly establish their formulation of quantum mechanics in configuration space. These books also discuss probability density and conserved probability current as entities that propagate in configuration space. All follow Born's derivation and (erroneously) ignore the fact that Green's theorem does not apply in configuration space, i.e. they all ignore Eq. (3.47). None of these textbooks ever describes what is meant conceptually by the concept of "wave propagation in configuration space". None of them ever fully define what is meant by the concept of a "*vector field ... in a configuration space ... with any number of dimensions.*" Also recall the section "Born's Ambiguity's Misuse by the Lab-Space Formulation School" above, wherein it is noted that Dicke and Wittke (1960) and Merzbacher (1961, 1970) both admit that these ideas make no sense at all for N > 1 particle systems. It would seem evident that if a wave cannot propagate in configuration space (as Dicke and Wittke and Merzbacher note), it is then difficult to understand how a current can flow in configuration space.

Landau and Lifshitz (1958, 1965, pp. 55–58) also claim (without comment) that a conserved probability current somehow propagates and/or flows in configuration space, without their giving any explanation of what this actually means. Bjorken and Drell (1964, pp. 2–9) derive a conserved probability current in configuration space, and use it to demonstrate flux conservation for the single-particle Dirac's equation. They conclude *"Integrating (1.20) over all space and using Green's theorem, we find $\partial/\partial t \int d^3x \ \psi^\dagger\psi = 0$, (1.23), which encourages the tentative interpretation of $\rho = \psi^*_{config}\psi_{config}$ as a positive definite probability density.* " They do not seem to notice the incongruity of their simultaneous claims that while Ψ_{config} *"has no direct physical interpretation, ...the probability interpretation requires that the sum of positive contributions to* $|\Psi_{config}|^2$ *for all values of* $(q_{1,config}, ..., s_{n,config})$ *at time* t *to be finite for physically acceptable wave functions* Ψ_{config}*."* How can a wave function have "no direct physical interpretation, but simultaneously be "*physically acceptable*"?

Additional confusion that arises from Born's ambiguity is described in the section "Some Observations Regarding Which School is Proper".

Quantum Field Theory 1—Quantization of Known Classical Fields

The third school of thought used for formulating quantum mechanics is what is called quantum field theory. It may be divided into two forms. The first form, herein called Quantum Field Theory 1, quantizes known real classical fields like light and sound, formulated in lab space for massless particles, that describe known *real stuff in real space–time*. The second form second quantizes abstract fields like wave functions for matter-wave fields, formulated in configuration space for massive particles, and/or hypothetical pseudo-classical abstract fields with no observed counterpart in nature. The second form, herein called Quantum Field Theory 2, is discussed in the section "Quantum Field Theory 2—Second Quantization of Wave-Functions".

There are two immediate applications of Quantum Field Theory 1. The first is to light, electromagnetic radiation, and the electromagnetic field. The second is to sound and vibrational displacements of atoms in solids. A primary purpose of this quantization is to provide these fields with a particle-like character. For light, the particles are photons. For sound, the particles are phonons. The need for this quantization and their associated particles originates with Einstein (1917). He noted that the electromagnetic field needs to be reformulated (quantized), in order to allow its description in terms of particles, that he called *"directional radiation bundles"* (a.k.a. light quanta, a.k.a. corpuscles, a.k.a. photons). Einstein demonstrated that without these *directional radiation bundles,* thermal equilibrium cannot be maintained between the radiation field and a gas of molecules.

The (canonical) procedure for quantizing a classical system, such as a classical field, is well defined, and its application to a lab space-classical field is straightforward. The two well known classical fields, light and sound, are thus readily quantized using it. First, one defines the classical system's "degrees of freedom". For a classical field, space is divided into an infinite number of infinitesimal cells, and the field's values at each cells' lab-space position can be taken to be the fields' degrees of freedom. These values, in turn, can qualify as properties of classical "stuff" at the cell's position. Schiff (1955, p. 344) uses this method. Alternatively, Fermi (1932) expands the field's values using a Fourier series, and the series coefficients are then used for the field's degrees of freedom. Second, one uses the degrees of freedom to form an associated Hamiltonian that defines the total energy of the system. For the classical electromagnetic field, the total classical field energy is well known from electromagnetic theory. For sound waves in a solid, atomic vibrational displacements from rest form a field. Assuming the atoms to be harmonically bound spring-mass systems, the total classical vibrational energy (a.k.a. thermal energy) is also readily calculated. Once the Hamiltonian has been defined, it is used in Schrödinger's equation to calculate the field's quantized dynamics.

The quantization of the electromagnetic field, along with the quantum theory of radiation and quantum electrodynamics, were originally developed by Fermi, Dirac, Heisenberg and Pauli. The essential elements are presented in an excellent early review article by Fermi (1932), who uses the canonical procedure to quantize the electromagnetic field. We shall follow Fermi's (1932) discussion in the section "Quantum Theory of Radiation and Quantum Electrodynamics". Photons are seen to emerge from his formalism. The experimental demonstration that photons act like Einstein's *directional radiation bundles* is outlined in the section "Einstein's Need for Directional Radiation Bundles".

A scrutiny of Fermi's treatment reveals the existence of three very different fields that are sometimes confused with each other via Born's ambiguity. (Fermi does not confuse them!) The first of these fields is the classical field itself. The field's classical value is uniquely defined at every point in lab space. The second field is the quantized version of this field. It is displayed via Fermi's equations (3), (4) and (12) in lab space using quantized mode amplitudes, u_s. The field's mode amplitudes (or values) are used as the degrees of freedom of the field. There are an infinite number of them. A third, related, but very different field is in the form of a function formulated in an abstract vector space. It is the *Schrödinger function* for the system (field plus atom), defined in configuration space by Fermi's equations (35). Its value depends on the field's and atom's degrees of freedom. It is computed as the solution to Schrödinger's equation for the system (Fermi's equations (48)), and it displays the calculated quantum dynamics of the system. Fermi carefully uses the term "*Schrödinger function*" rather than the term "*wave function*" to prevent confusion between this function and the quantized classical field's functional dependence on position.

Fermi's *Schrödinger function's* arguments are in configuration space, to be distinguished from the quantized classical field function's argument, which is the position in lab space where the field may be evaluated. What does the Schrödinger function's *value* mean? Similarly to Schrödinger's so-called wave function in configuration space in standard quantum mechanics, when integrated over all (configuration) space, it gives the particle number density. When only one particle is present, this is the integrated probability density, and it equals one, which, of course, is now equal to the number of particles present. Unlike standard quantum mechanics, in a quantum field theory the number of particles may, however, change with time.

Fermi attempts to show that a causal, *real stuff in real space–time,* behavior for the electromagnetic field is maintained for his quantized field. Unfortunately, Fermi fails to notice that quantization in configuration space brings in the possibility of entanglement, whereupon entanglement of separated photons ruins any hopes for such a result. When entanglement is present, measurements of the field at widely separated positions in lab space unfortunately destroy any residual hopes for a causal, *real stuff in real space–time,* behavior for the quantized electromagnetic field. This fact is quite the opposite of the claims made by Bohr and Rosenfeld as outlined by Heitler (1954, 1957, 1960, pp. 76–86), who consider only the unitary evolution of the electromagnetic field, and do not consider its non-unitary evolution that occurs via von Neumann's collapse process. This important fact was first demonstrated by experimental tests of Local Realism, first performed by Freedman and Clauser (1972). More recent experiments by Gisin (2002) starkly demonstrate the extreme lack of causal unitary evolution for the quantized electromagnetic field that occurs via von Neumann's collapse process. This latter important behavior is discussed below in the section "von Neumann's Collapse of the Entangled Two-Photon Quantized Electromagnetic Field".

Following Fermi's (1932) treatment of the subject, many textbooks have been written and improvements to the theory have been added. They are discussed below in section "Improvements to Fermi's treatment of field quantization". Quantization of sound waves and the emergence of phonons from the formalism is described by Henley and Thirring (1962) and by Kittel (1953, 1956). It is performed similarly to Fermi's procedure for the electromagnetic field, and is not discussed here.

Quantum Theory of Radiation and Quantum Electrodynamics

In Fermi's (1932) description, the quantization of the classical electromagnetic field consists of two parts. The first part quantizes the radiation field. Fermi calls this part of the theory the *"quantum theory of radiation"*. In the second part, Fermi attempts to quantize an electromagnetic field of *"the most general type that cannot be constructed by simply superposing plane electromagnetic waves"*. Fermi calls this second part of the theory *"quantum electrodynamics"*,

Fermi's treatment of the quantum theory of radiation begins by using the values of the electromagnetic vector potential $A_{lab}(r_{lab}, t)$ at all points r_{lab} as the basic degrees of freedom of the system to be quantized. He expands A_{lab} in terms of a set of Fourier-series standing-wave field modes that are functions of r_{lab}. The modes are assumed to exist in a very large rectangular cavity. At the end of the solution, the walls of the cavity are allowed to expand to infinity.[20] The resulting Fourier transform then transforms the spatial degrees of freedom of the field to allow the Fourier coefficients for these modes to be the new transformed degrees of freedom.

In Fermi's treatment, Maxwell's equations describe the classical electromagnetic field in lab space. The energy density of the electromagnetic field at every point in lab space is defined via Maxwell's equations and the Lorentz force law, consistently with Lagrange's and Hamilton's laws of motion. The total energy integrated over all of lab space is correspondingly used to define a Hamiltonian function in terms of the field's mode amplitudes. A Schrödinger's equation (in matrix form) for the evolution of the mode amplitudes is then formulated. Its solution yield's the *Schrödinger function* that, in turn, displays the system's resulting dynamics. The mode energies are all found to act like the excitations of a simple harmonic oscillator. Quantum mechanically, the simple harmonic oscillator has well-known solutions, consisting of a set of states, whose energy levels are equally spaced by energy intervals, $\hbar\omega$, where ω is the field mode's angular frequency. Each incremental excitation is then *interpreted* to consist of the presence within the quantized electromagnetic field of a particle-like photon[21] with energy $\hbar\omega$. To handle atom–field interactions, Fermi couples the field-mode degrees of freedom to the configuration-space degrees of freedom of one or more atoms. Thus, when the atom's degrees of freedom and Hamiltonian are added to that of the radiation field, he obtains *"the fundamental equation of the radiation theory"*.

Fermi attempts to show that his formalism for the quantized electromagnetic field describes *real stuff* in *real space–time* by showing that it provides reasonable predictions for five important effects. He considers (1) Emission from an excited atom; (2) Propagation of light in a vacuum; (3) A case of interference—the Lippman friges, (4) The Doppler effect, and (5) The Compton effect. Fermi's demonstration of *"Propagation of light in a vacuum"* is particularly interesting. In it, he demonstrates the self-consistency of his formalism with regard to the notion of a wave front propagating at light speed between the two widely separated atoms. Both atoms are coupled to the standing-wave modes of the now infinitely wide radiation field cavity modes. One of the atoms is initially excited, and the other is not. He shows that, following the decay of the first atom, the atom's energy is first transferred to an excitation of a phased-superposition of single photon modes of the radiation field. Fermi then shows that because of this carefully phased-superposition, thereafter the second atom cannot be detected in an excited state until a speed of light travel time between

---

[20] A problem with Fermi's method that does not seem to have been addressed by him at the time, is that no mirrored box cavity is present in most labs. Indeed, the interiors of most laboratory apparatuses used to test quantum electrodynamics have light absorbing walls, and not perfectly reflecting walls. However, when experiments are actually performed mirrored box cavity, a surprising new wealth of physics is uncovered. See, Berman (1994), A theory with absorbing walls does not appear to have been formulated.

[21] Concerning the use of the term *"interpret "*, note a big jump in logic that occurs here without experimental justification, especially if the photon, hereby defined, is to have the same particle-like properties required by Einstein's (1917) "directional radiation bundles." (See the section "Einstein's Need for Directional Radiation Bundles".).

the two atoms has elapsed. Fermi thereby shows that the radiation field, which is now comprised of a sum of an appropriately-phased set of infinitely-wide standing waves, can produce the effects of a traveling wave "photon" with light-speed causal effects. This important effect is true, however, only for an $N = 1$ particle excitation (single photon) of the electromagnetic field. Unfortunately, Fermi did not consider an $N \geq 2$ particle excitation of the field, where the hoped for causal effect fails miserably. (See the section below "von Neumann's Collapse of the Entangled Two-Photon Quantized Electromagnetic Field".)

Einstein's Need for Directional Radiation Bundles

Einstein (1917) in his seminal paper "*On the quantum theory of radiation*" introduced the concept of particle-like "photons", that he called "*directional radiation bundles*". In that paper, he considered a gas of molecules interacting with electromagnetic radiation (light), and derived the necessary conditions for thermal equilibrium to be maintained between them, and for the second law of thermodynamics to hold in the face of quantum theory. He pointed out that these *directional radiation bundles* (a.k.a. field quanta, a.k.a. corpuscles, a.k.a. photons) must exist. These *directional radiation bundles* must be emitted and absorbed (a.k.a. created and annihilated) by molecules in a gas totally at random. They must conserve energy and momentum in not two but three important processes—absorption, emission, and stimulated emission. He derived the necessary rate coefficients for these three processes. They are now known as the Einstein A and B coefficients. Einstein thus showed that the classical electromagnetic field must be "*quantized*" in terms of these particle-like bundles.

Einstein's need for these "*directional radiation bundles*" is, in fact, a primary motivation for quantizing the electromagnetic field. Correspondingly, it is a little surprising that a particle like behavior for photons is not discussed at any significant length by Fermi, although he does demonstrate conservation of momentum and energy between the atoms and the field. In this absence, there remained an outstanding need to address the issue of Einstein's required particle-like behavior for the photons. Need for an experimental confirmation of this behavior was further noted by Schrödinger (1927) [see also, Jauch (1971)]. Schrödinger pointed out that that a particle-like photon (a directional radiation bundle) will be either reflected or transmitted at a half-silvered mirror (as required by Einstein, and by the quantum theory of radiation), while a wave-like directional radiation bundle, if governed solely by Maxwell's equations, will be both simultaneously reflected and transmitted at a half-silvered mirror. Schrödinger proposed that an experiment should be performed to find out which of these two predictions is true. A first attempt by Ádám, Jánnosy, and Varga (1955) to test Schrödinger's idea was inconclusive, and resulted in only a null experiment, although this fact was not noticed at the time. A fully conclusive experimental demonstration of Einstein's required behavior was finally performed by Clauser (1974), wherein single photons are indeed observed either to be reflected or to be transmitted at a half-silvered mirror, but not both transmitted <u>and</u> reflected simultaneously.

von Neumann's Collapse of the Entangled Two-Photon Quantized Electromagnetic Field

Fermi's two-atom excitation demonstration of *"Propagation of light in a vacuum"* was intended by him to reveal a reasonable causal behavior of the quantized electromagnetic field. His argument works very "nicely" for single-photon excitations of the quantized electromagnetic field. On the other hand, the definition of the word "nicely" used here depends upon one's point of view regarding whether or not a quantized electromagnetic field retains its ability to describe *real stuff in real space–time*. In Fermi's two atom excitation scheme, a detection of the second atom in its excited state immediately precipitates a von Neumann collapse of the spatially-extended quantized electromagnetic field and of its carefully phased standing wave modes. These modes are infinitely-wide and defined everywhere in space. Nonetheless, they collapse instantaneously everywhere in space to an un-excited zero-photon excited state. Unlike the (perhaps discernable) traveling wave-front in lab space that originally causally induces the second atom's excitation, there is definitely no discernable traveling wave-front in lab space that describes the field's de-excitation via the von Neumann collapse. Unfortunately, the collapse's only observable effect is the field's loss of any ability to perform subsequent other atomic excitation.

So, no harm equals no foul? Yes foul! Fermi's argument definitely does not extend "nicely" for two-photon excitations of the infinitely-wide spatially-extended quantized electromagnetic field. Tests of the CHSH and CH inequalities most often employ two-photon polarization-entangled-state excitations of the field. Consider what happens to such a two-photon excitation under the assumption that the quantized radiation field is assumed to represent CH's *real stuff in real space–time*. When the field polarization of one particle-like photon component of the pair is measured, the other particle-like photon component's polarized field is instantaneously collapsed via von Neumann's collapse process of the quantum state of the field. It is instantly reset everywhere to match the polarization of the measured polarization of the first component photon. This collapse occurs instantaneously without Fermi's light travel time delay having elapsed. Unlike Fermi's causal speed-of-light behavior of the quantized field, von Neumann's collapse process instead occurs instantaneously, with an apparently <u>infinite</u> speed. Following a measurement of the polarization of the first photon's field, immediately thereafter the second photon's field can then be detected in a polarization-parallel state.

Lack of nicety gets worse! Gisin (2002) speculated that von Neumann's collapse process might be a *real causal process in real space–time* within the quantized electromagnetic field, and that a collapse-wave-front correspondingly propagates causally in *real space–time*, with a speed perhaps at or less than that of light. However, he importantly noted that if von Neumann's collapse is assumed to be a *real process*, its behavior is not Lorentz invariant. Thus, the possibly finite speed of any collapse-wave-front depends upon what absolute reference frame the experiment is performed in. Gisin tested his hypothesis and has experimentally set a lower limit to the speed of the collapse-wave-front to be 2/3 10^7 and 3/2 10^4 times the speed of light, depending on whether the choice of reference frame is taken to be that of the local "Swiss Alps", or that of the cosmic background radiation.

Gisin (2002) also noticed that if von Neumann's collapse process is assumed to be a *real process*, it is possible to build an apparatus pair with each apparatus moving with respect to the other and with respect to the two-photon source. Each apparatus then measures the polarization of its member photon from an entangled photon pair. The apparatus motions are carefully designed so that each measurement occurs before the other measurement takes place. Zounds! Neither measurement can then precipitate a von Neumann collapse of the field if collapse is a causal process. Gisin thus notes that *"If each measurement happens before the other, then the quantum correlation should disappear, however large the speed of the spooky action!"* He performed the experiment and observed that the quantum correlation persists! Double zounds. Von Neumann's collapse of the quantized electromagnetic field that occurs because of its being measured really is spooky! But everyone knows that ghosts are not real, don't they? The conclusions that one can draw from Gisin's experiments are that, von Neumann's collapse of the quantized electromagnetic field cannot be viewed as a real causal process! The second conclusion is that, despite Fermi's attempt to demonstrate that a quantization of the electromagnetic field allows it to retain some semblance of a description of CH's *real stuff in real space–time*, is a total flop when field measurements are made at remotely separated points in space. Quantization of the field definitely renders it no longer describable as *"real stuff in real space–time"*, and photons, in particular, are also not describable as CH's *real stuff in real space–time*. Of course, that same conclusion was obtained earlier from experiments that test the CH and CHSH inequalities.

Improvements to Fermi's Treatment of Field Quantization

There remain additional outstanding problems with Quantum Field Theory 1. Although some have improved over time. Fermi does show that the quantum theory of radiation is sufficient to explain many of the known experimental results to date. On the other hand, he does not show that field quantization is necessary to explain these experimental results. In the early 1970s, said necessity was called into question by many workers in the field of quantum optics, when many experimental results were found to be readily explained in terms of semi-classical radiation theory. Freedman and Clauser (1972) showed that experiments that measured the polarization correlation of entangled two-photon states did indeed demonstrate the necessity. Clauser's (1974) additional experiments further demonstrate a need for field quantization.

Additionally, Fermi notes that his formalism is not without difficulties. He observes that the second part of the theory, quantum electrodynamics, "...*runs into serious difficulty ... every charge has an infinite electrostatic self-energy.*" That difficulty echoes back to his use of perturbation theory while addressing the first part of the theory, the quantum theory of radiation. At the end, Fermi admits "*To all these difficulties no satisfactory answer has been given.*" Fermi's method for addressing quantum electrodynamics was summarized subsequently by Feynman (1962, pp. 3–4). Thirty year later, Feynman commented that Fermi's difficulties (with infinite self-energies) leads to "*one of the central problems of modern quantum electrodynamics.*"

Fermi's treatment of field quantization was chosen for analysis here because it allows a graphic display of the existence of two very different fields,—an abstract vector-space "field", the *Schrödinger function*, whose arguments are in configuration space, that is to be distinguished from the quantized classical field, whose function's arguments indicate the position in lab space where it may be evaluated. This choice is not meant to downplay dramatic advances made in the field of quantum electrodynamics that were made in the late 1940s. The reader is directed to the collection of papers on the subject in the volume Selected Papers on Quantum Electrodynamics, Schwinger (1958) for more details on the subject. Quantization of the electromagnetic field is also described in the latter chapters of some of the quantum mechanics textbooks discussed above, e.g. Schiff (1955, Chaps. XIII–XIV), Merzbacher, (1961,1970, Chaps. 20–22), Messiah (1961, Chap. XXI), as well as in other textbooks that are devoted entirely to the subject, e.g., Bjorken and Drell (1965), Harris (1972) and Heitler (1954,1957,1960).

Some Observations Regarding Which School is "Proper"

Born's argument space ambiguity, once revealed, provides greater insight regarding the differences between Bohr and Einstein in their debate regarding the Einstein Podolsky Rosen (1935) "EPR" paradox. They now appear to have been simply talking past each other, each assuming a different argument space for a two-particle wave function, without recognizing this fact. Bohr insisted that a measurement of one particle of an entangled pair disturbs the whole two-particle (global) wave function, no matter how far apart the particles are. Since there is no space–time description for this global wave function, there is no causality problem associated with the "disturbance" of the kind addressed by Gisin (2002). Bohr was presumably assuming a configuration-space formulated wave function. Configuration-space is a purely abstract mathematical entity—a mathematical (sometimes infinite-dimensional) vector space. As we have noted above, it has no dependence on lab space through which a causal disturbance (e.g. von Neumann's) might propagate. Einstein's thinking was evidently conceptually anchored in lab space. His concept of a wave function was presumably that of $\Psi_{lab}(\mathbf{r}_{lab}, t)$, i.e. one formulated in lab space. In such case, any such causal disturbance (e.g. von Neumann's) must obviously propagate as a wave through lab-space. By contrast, in Bohr's case, Ψ_{config} depends only on configuration-space variables, and thus has the same value everywhere in lab space. Its very nature is <u>inherently</u> non-local. Super-luminal wave propagation in configuration-space is not prohibited, since it is totally divorced from lab space. Indeed, as noted above, there is no wave propagation at all, let alone any wave propagation from one apparatus to the other, whereupon non-local wave-function collapse is not inconsistent with special relativity. This fact is true no matter how many particles are being described by Ψ, but it only becomes conspicuous and bothersome when at least two or more widely separated particles are being described by Ψ_{config}, i.e. when its spatially constant value depends on the dynamics of both of these separated particles.

As a result of Born's ambiguity and the two schools of thought that emanate from it, we have thus identified two very different, indeed incompatible, formulations of wave functions that are used by practitioners of the art. Following von Neumann's and Messiah's stern warnings, and following the observation here that lab-space formulations are evidently refuted by experiment, we shall define wave functions that are formulated in lab space as "improperly" formulated, and those formulated in configuration space will be defined here as "properly" formulated.[22] Wave functions are also sometimes called" first-quantized fields, whether or not they were "properly" formulated. Wave functions (in standard quantum mechanics, but not in quantum field theory) conserve particle number, independently of what space they are formulated in. Indeed, Born's conserved probability current (improperly formulated in lab space) is constructed from these first-quantized improperly (lab-space) formulated fields. Indeed, it is the basic purpose of Born's current to ensure particle conservation, as was discussed above in the section "Born's Probability Density and Conserved Probability Current Defined in Lab Space". We have also noted above that Born's ambiguity is sometimes improperly used to interchange the meanings of these fields. When the quantum field theory of the matter-wave field is discussed further in the section "Quantum Field Theory 2—Second Quantization of Wave-Functions", said improper use of Born's ambiguity will be seen to get much worse. Classical fields, as defined in the section "Laboratory Space and Classical Fields" above, are by their very nature, definable "properly" only in lab space.

A revealing of Born's argument space ambiguity also spotlights von Neumann's wave function collapse. For such a collapse to make any semblance of sense, it appears that wave functions must be formulated "properly" in configuration space, as von Neumann and Messiah insist. Said collapse occurs everywhere in space–time when a single particle is detected at a specific point in space–time. Just as with the EPR collapse of a two-particle wave function, a single particle configuration-space wave function must instantaneously collapse when, or even in anticipation of the other particles measurement. Given that the collapse occurs only in an abstract configuration space, there is no need for a super-luminal wave collapse to propagate throughout lab space. *Super-luminal wave propagation in configuration-space is not*

---

[22] Also, it is noted above that while lab-space formulated wave functions attempt to describe Local Realism's *real stuff in real space–time*, they cannot do so in general, and must therefore be considered *improper*ly formulated.

prohibited in configuration space. Gisin's (2002) experiments demonstrate that fact with stark clarity. The fact that it must collapse in anticipation of a measurement is particularly difficult to understand, only if it is to be viewed as a real process in (lab) space–time.

Abstract configuration space is thus inherently inscrutable by its very nature, as the Copenhagen "interpretation" of quantum mechanics has long professed. It thus allows magic to happen! In a configuration-space formulation of quantum mechanics, no objects are needed to exist as "stuff" in lab space. As seems to happen in stage magic, objects can simply appear and disappear. Since there is no stuff needed to be objectively present in lab space, since there are no objects, then faster than light signaling between non-existent objects, and action at a distance between non-existent objects is not issue. Since the configuration-space wave function does not represent a classical field (as defined above in the section "Laboratory Space and Classical Fields"), its value is not a function of lab space position, and the wave function, Ψ_{config}, itself, has no evident spatial dependence. Correspondingly, a non-causal non-unitary collapse that results from a measurement operation is no longer a conceptual problem. In fact, there are no conceptual problems, because there are no conceptual physical models to deal with. Instead, there are only abstract mathematical concepts to deal with. It seems that one must accept the magic, however distasteful that may be!

Quantum Field Theory 2—Second Quantization of Wave-Functions[23]

In the above section "Quantum Field Theory 1—Quantization of Known Classical Fields", we divided quantum field theory into two forms. The first form quantizes known real classical fields that describe known *real stuff in real space–time*, i.e. stuff with real readily observed classical components—light and sound. The second form second quantizes abstract fields like wave functions for matter-wave fields, and/or hypothetical pseudo-classical abstract fields with no observed counterpart in nature. A basic purpose of field quantization is to unify the description of (first) quantized classical fields and second-quantized wave functions, in such a way that allows a varying number of particles to be somehow associated with these various and diverse fields, and to do so in a manner that allows particle creation and annihilation.

---

[23] Definitions of the terms *"first and second quantization"* as used here are those given by Schiff (1955, p. 348) in his section 46 *"Quantization of the nonrelativistic Schrödinger equation"*. He says *"This application implies that we are treating Eq. (6.16)as though it were a classical equation that describes the motion of some kind of material fluid. As we shall see, the resulting quantized field theory is equivalent to a many particle Schrodinger's equation, like (16.1) or (32.1). For this reason, field quantization is often called second quantization; this term implies that the transition from classical particle mechanics to Eq. (6.16)constitutes the first quantization."* Bjorken and Drell (1965, Chap. 13) similarly use the term in "Second Quantization *of the Dirac equation*", while they discuss in their Chap. 14. *"Quantization of the electromagnetic field"*.

Unfortunately, it appears that said unification can occur only with a generous use of Born's ambiguity, given the wide variety of fields that various authors try to sweep together.

In the section "Quantum Theory of Radiation and Quantum Electrodynamics" we described how Fermi quantized a known classical field, the electromagnetic field, in order to produce photons. Doing so, he prescribed the need for two very different field types—those whose arguments are defined in configuration space, and those whose arguments are defined in lab space. The first is the abstract vector-space function that he named the *Schrödinger function*. Its arguments are in configuration space. It is clearly not a classical field, as defined above in the section "Laboratory Space and Classical Fields". It is to be distinguished from the quantized and non-quantized classical electromagnetic fields, whose function's arguments indicate the position in lab space where they may be evaluated. These very different field types are sometimes confused with each other via Born's ambiguity. Indeed, they have a direct parallel to the two schools of thought we have identified and distinguished in the sections "The lab-space formulation of quantum mechanics" and "The Configuration-Space Formulation of Quantum Mechanics". The primary distinction between them is in regards to their associated argument spaces.

There at least 7 mathematically distinct kinds of fields noticed in the reviewed literature on quantum field theory. Some are considered worthy of consideration for field quantization by various authors. Some are actual classical fields that describe known *real stuff in real space–time*. Some are abstract functions in a vector space. Some are totally hypothetical in nature. Following the above definition of "proper", it is noteworthy that some are "properly" defined, and others are "improperly" defined. They include the following fields:

1. Classical fields defined properly only in lab space. These include the classical electromagnetic field that gives rise classically to light waves and quantum mechanically to photons, and the field of vibrational displacements that gives rise classically to sound waves and quantum mechanically to phonons. When acting classically, these fields can host real wave motion in lab space.

2. Quantized version of field #1.(quantized as per Fermi (1932). It is defined properly only in lab space. It can undergo non-causal non-unitary von Neumann collapse. Case 1 fields cannot similarly collapse.

3. Fermi's "*Schrödinger function*" for case #2 quantized atom(s) plus field. It is a vector-space function defined properly only in the configuration space of the atom or atoms and the field degrees of freedom (as per section "Quantum Theory of Radiation and Quantum Electrodynamics"). A *Schrödinger function* is similarly definable for the atom or atoms alone and for the field alone when these entities do not interact.

4. Single-particle wave function (a.k.a. the matter-wave field) defined improperly in lab space for a single particle, that is a solution to the lab-space improperly defined Schrödinger, Klein Gordon and/or Dirac equations. It is the field depicted in Figs. 3.1–3.7. It cannot be properly generalized to describe N particle

systems (as per the section "Quantum Theory of Radiation and Quantum Electrodynamics"), although Schiff (1955, p. 341) claims that field quantization provides a proper method for doing so.[24] Henley and Thirring (1962), p. 3) also consider such fields as suitable for field quantization.

5. Pseudo-classical,[25] "classical free field", similar to a case 4 field, that satisfies, say, the Klein–Gordon equation. Such fields have no observed counterpart in nature. The field is defined in lab space, although it is totally ambiguous as to whether this definition is proper or improper. It is used by various authors, e.g. by Bjorken and Drell (1965), and Messiah (1961, p. 960), to demonstrate methodology for second quantizing fields with no known classical counterpart, whereupon there is no classical way to calculate the associated energy density. Bjorken and Drell (1965, p. 34) note that "*Classically, the field $\varphi(x)$ is observable and its strength at a point x can be measured.*" Many authors further consider functions and define pseudo-classical fields that are typically permanently complex and/or have non-geometrical abstract properties, such as Dirac spinors. Sometimes, authors quantize a "real scalar field" and seem to play ambiguously on words. They use the word "real" to mean real-valued, i.e. having no imaginary component, rather than known to really exist in nature. (Are ghosts real or imaginary, or even complex?) Various authors thus stretch the limits of one's imagination regarding what may be considered "classical", and/or "real".

6. Single-particle wave function (a.k.a. the matter-wave field) defined only properly in configuration space. It is related to the N particle wave function using a sum of products form, as per sections "Factorization of Schrödinger's N-particle Configuration-Space Wave Function" and "Wave-Function Factorization or Not!". In actuality, it is already a case 3 "Schrödinger function", as per Fermi's definition and usage.

7. N-particle wave function (a.k.a. the matter-wave field), defined only properly in configuration space. In actuality, it is a "Schrödinger function" as per Fermi's definition and usage. Recall that Bjorken and Drell (1964, p. 2) state that properly defined vector-space wave functions that are used as N-particle wave functions have *no direct physical interpretation*. Nonetheless, Bjorken and Drell (1965, Chap. 13) second quantize them (but only when the particles are Fermions).

Case 1 quantization of both light waves and sound waves, properly defined in lab space, discussed above in the section "Quantum Field Theory 1—Quantization of Known Classical Fields", seems to make sense. First quantization of particle motion, properly defined in configuration space, as discussed above in the section

---

[24] Schiff (1955, p. 341) says "*The field quantization technique can also be applied to a Ψ field, such as that described by the non-relativistic Schrödinger Eq. (6.16)or by one of the relativistic Eqs. (42.4)or (43.3). As we shall see (Sec.46), this converts a one-particle theory into a many particle theory; ... Because of this equivalence, it might seem that the quantization of the fields merely provides another formal approach to the many-particle problem. However, the new formalism can also deal with processes that involve the creation and destruction of material particles.*"

[25] Let's pretend that such a classical field exists although we have no evidence that it does.

"The Configuration-Space Formulation of Quantum Mechanics", also perhaps makes sense. It is totally <u>ambiguous</u> as to whether Case 5 first quantization of pseudo-classical fields makes sense, given their vague definition. But, as we have argued in the section "Quantum Field Theory 1—Quantization of Known Classical Fields", a configuration-space wave function is not a classical field in lab space, otherwise it cannot undergo von Neumann collapse. Thus, it seems totally <u>improper</u> to second quantize case 6 and case 7 abstract configuration-space wave functions. Indeed, the only means to do so appears to be to use Born's ambiguity to convert these to be case 4 wave functions or case 5 pseudo-classical wave functions, <u>improperly</u> defined in lab space, and then to proceed by pretending them to be classical fields. Born's ambiguity correspondingly appears to be an integral foundation of quantum field theory of the matter-wave field. Unfortunately, it now appears also to be a major impediment.

If the goals of classical-field quantization and second quantization of wave-functions are to describe real-stuff in real space time, as Fermi apparently has tried to do with the classical electromagnetic field, in terms of a variable number of spontaneously forming particles, then that goal evidently cannot be achieved for any kind of field. It seems that the best that one can hope for is to describe totally abstract stuff (including pseudo classical fields) in real space time with a variable number of spontaneously forming particles. The new remaining problem identified here is how is one to resolve Born's ambiguity *vis à vis* second quantization of mater-wave wave functions.

Conclusions

What then does quantum mechanics <u>not</u> describe? Despite pretenses made in various quantum mechanics books, quantum mechanics does not describe the dynamics of any thing that is objectively real. Have we gotten any closer to answering the question posed in the opening paragraph of this paper?—What do standard quantum mechanics and quantum field theory describe? It seems that the answer is—Not really. What quantum mechanics and quantum field theory seem to describe is something other than objectively real stuff evolving in space–time. It seems that the best that an *improperly* (and inconsistently) formulated form of quantum field theory for matter-waves can offer is a description of totally abstract stuff (including pseudo-classical fields) that evolves non-causally in real space time with perhaps a variable number of spontaneously forming particles.

The basic conclusion obtained from a consideration of the experimental refutation of Bell–Clauser–Horne–Shimony Local Realism is that any theory that describes *real stuff* evolving causally in *real space–time,* must violate experiment for its predictions for separated entangled systems, assuming that it is capable of making predictions for these systems. Lab-space formulations of standard quantum mechanics are unable to make the necessary predictions. Lab space formulations of standard quantum mechanics qualify as theories of Local Realism; however, such formulations of quantum mechanics are unable to make the necessary predictions, thereby to reveal their own "dirty laundry". In any case, lab-space formulated standard quantum mechanics provides, at best, an *"improper"* description for matter waves, where "propriety" instead requires the formulation in configuration space only.

A dilemma is thus revealed by this article—how does one find an *"unambiguous"* and *"proper"* method within standard quantum mechanics and quantum field theory to provide a variable particle number for massive particles that have no known associated classical field. To be *"unambiguous"*, the method must avoid use of Born's ambiguity, which, in turn, erroneously considers configuration space and lab space to be equivalent and interchangeable. Unfortunately, at present, Born's ambiguity appears to be an integral foundation of quantum field theory of the matter-wave field for massive particles.

Related questions are—can we live with Born's ambiguity, and/or equivalently, can we experimentally distinguish between the different lab-space and configuration-space formulations of quantum mechanics and quantum field theory? That is, can we have it both ways? It seems that we cannot. The configuration space formulation seems to be required, as per the experimental refutation of Bell–Clauser–Horne–Shimony Local Realism. The answer leaves open the remaining question identified here—How does one resolve Born's ambiguity *vis à vis* second quantization of matter-wave wave functions.

Sadly, for those of us who had hoped to find a theory of *real stuff in real space–time*, we are left with the situation wherein quantum mechanics describes abstract, impossible-to-be-real stuff. Said stuff is described by an abstract mathematical framework, wherein said stuff somehow moves and propagates perhaps stochastically (if the words "moves" and "propagates" still retain any understandable meaning) in an abstract multi-dimensional mathematical space that may contain non-classical abstract components. Whatever quantum mechanics does indeed describe is sadly very difficult to visualize.

There is one final perhaps important irony regarding the *"demise"* of any theory that attempts to describe *real stuff in real space–time*. Said *demise* now includes lab-space formulations of quantum mechanics and quantum field theory. Both Albert Einstein and Max Born[26] were a strong proponents of the quest for a lab-space description (i.e. a. space–time description) of natural phenomena. Correspondingly, said *demise* must certainly have an impact on one's conceptual understanding of Einstein's theory of general relativity, if and when it is reconciled with quantum mechanics. General relativity certainly seems to be a theory within the bounds of Local Realism. Indeed, a fundamental tenant of general relativity is that the geometry of space–time depends on the mass energy content of how much *real stuff* is present

---

[26] This article has concentrated on Born's quest for a space–time description of natural phenomena. It is noteworthy that the first printing of Born's (1933) seminal textbook predates the famous "EPR" paper by Einstein, Podolsky, and Rosen (1935).

in *real space–time*. Similarly, Stephen Hawking's claims that information is always conserved and is contained in an accessible volume of space–time, especially at a black-hole boundary, seems to be based a use of Born's conserved probability current, which we have noted above is not really conserved in the real space–time framework of general relativity.

References

Ádám A, Jánnosy L, Varga P (1955) Acta Phys Hung 4:301; Ann Physik 15:408

Apostol T (1961) Calculus. Blaisdell Publishing Company

Bethe H, Salpeter E (1957) Quantum mechanics of one- and two-electron atoms. Springer-Verlag

Bell J (1964) Physics 1:195

Bell J, Shimony A, Horne M, Clauser J (1976–1977) Epistemological letters (Association Ferdinand Gonseth, Institut de la Methode, Case Postale 1081, CH-2501, Bienne.) republished as An exchange on local beables. Dialectica 39:85–110 (1985)

Berman P (1994) Cavity quantum electrodynamics. Academic Press, Inc

Bjorken J, Drell S (1964) Relativistic quantum mechanics. McGraw Hill

Bjorken J, Drell S (1965) Relativistic quantum fields. McGraw Hill

Born M (1933, 1935, 1969) Moderne Physik. References herein are to the 1969, 8th Edition English translation. It was renamed *Atomic Physics* because its publisher, Blackie and Son, already had an English language textbook named Modern Physics

Born M, Wolf E (1959, 1987) Principles of optics. Pergamon Press

Clauser J, Horne M, Shimony A, Holt R (1969) Phys Rev Lett 23:880–884 (commonly referred to as CHSH)

Clauser J (1974) Phys Rev D 9:853–860

Clauser J, Horne M (1974) Phys Rev D 10:526–535 (commonly referred to as CH)

Clauser J (1976) Phys Rev Lett 36:1223

Clauser J, Shimony A (1978) Bells' theorem: experimental tests and implications. Rep Prog Phys 41:1881–1927

Clauser J (2017) Bell's theorem, Bell inequalities, and the "probability normalization loophole". Chapter 28. In: Bertlmann R, Zeilinger A (eds) Quantum [un]speakables II, half a century of Bell's theorem. Springer International, Switzerland

Condon E, Shortley G (1964) The theory of atomic spectra. Cambridge University Press

Dicke R, Wittke J (1960) Introduction to quantum mechanics. Addison-Wessley

Dirac P (1930, 1935, 1947) The principles of quantum mechanics. Oxford at the Clarenden Press

Einstein A (1917) Phys Zs 18:121; English translation: On the quantum theory of radiation. In: van der Waerden B (1967) Sources of quantum mechanics. Dover Publications, Inc

Einstein A, Podolsky B, Rosen N (1935) Phys Rev 47:777–780

Eisle J (1964) Advanced quantum mechanics and particle physics from an elementary approach. The National Book Company. Taiwan

Eisberg R (1961, 1967) Fundamentals of modern Physics. Wiley

Fermi E (1932) Quantum theory of radiation. Revs Mod Phys 4:87

Feynman R (1948) A space-time approach to non-relativistic quantum mechanics. Rev Mod Phys 20:367

Feynman R (1962) Quantum electrodynamics. Benjamin Inc., W. A

Feynman R, Hibbs A (1965) Quantum mechanics and path integrals. McGraw Hill Book Company

Freedman S, Clauser J (1972) Phys Rev Lett 28:938–941

French A, Taylor E (1978) An introduction to quantum physics. W. W. Norton and Company

Gisin N (2002) Sundays in a quantum engineer's life. chapter 13, In: Bertlmann R, Zeilinger A (eds) Quantum [un]speakables, from Bell to quantum information, Proceedings of the 1st quantum [un]speakables conference. Springer International, Switzerland

Goldstein H (1950) Classical mechanics. Addison-Wessley, Reading MA

Harris E (1972) A pedestrian approach to quantum field theory. Wiley

Heitler W (1954, 1957, 1960) The quantum theory of radiation, Oxford at the Clarendon Press

Henley E, Thirring W (1962) Elementary quantum field theory. McGraw Hill Book Company

Jauch J (1971) Are quanta real? Indiana University Press, Bloomington Indiana

Kittel C (1953, 1956) Introduction to solid state physics. Wiley

Landau L, Lifshitz E (1958, 1965) Quantum mechanics, non-relativistic theory. Pergamon Press

Margenau H, Murphy G (1943, 1956) The Mathematics of Physics and Chemistry. D. Van Nostrand Company

Merzbacher E (1961, 1970) Quantum mechanics. Wiley

Messiah A (1961) Quantum mechanics. North Holland Publishing Co., Amsterdam

Pauling L, Wilson E (1935) An introduction to quantum mechanics. McGraw Hill Book Company

Public Broadcasting System (2018) Einstein's quantum riddle. Nova television documentary

Schiff L (1955) Quantum mechanics. McGraw Hill Book Company

Schwinger J (1958) Selected papers on quantum electrodynamics. Dover Publications

Schrödinger E (1927) Ann Physik 82, 124, English translation in Schrödinger E (1928) *Collected papers on wave mechanics.* Blackie & Son, Ltd

Spitzer L (1956) Physics of fully ionized gases. Interscience Publishers

von Neumann J (1932) Mathematiche Grundlagen der Quantenmechanik. References herein are to the English translation, von Neumann J (1955) mathematical foundations of quantum mechanics. translated by R. Beyer, Princeton Univ. Press

Chapter 4
Neutron Spin Pendellösung Resonance

Kenneth Finkelstein

Introduction

When a particle with spin magnetic moment is scattered by a system with an electric charge distribution, spin–orbit interaction contributes to the scattered intensity. Schwinger (Schwinger 1948) first proposed that spin–orbit (SO) interaction would play a part in neutron-nuclear scattering. The effect occurs along with nuclear scattering when the neutron travels through the atomic electric field and senses, by way of its moment, a magnetic field due to its motion. The field is proportional to v/c (v is the magnitude of neutron velocity and c is the speed of light), points normal to the plane of scattering, and changes sign from side to side across the nuclear center. This asymmetry contributes an imaginary scattering amplitude that can be substantial for fast moving, polarized particles as discussed by Fermi (1954). For thermal neutrons, the SO-scattering amplitude is about 1% of nuclear scattering (10^{-4} effect on intensity) as shown experimentally by Shull (1963) using polarized neutron scattering in vanadium crystals.

The perfect crystal diffraction phenomenon, pendellösung oscillation, was first discussed by Ewald (1916). He pointed out that solutions of equations describing dynamical diffraction of x-rays can be strikingly similar to normal mode solutions for coupled pendula in classical mechanics. Pendellösung are a periodic swapping of amplitude between forward and Bragg propagation directions, of x-rays or neutrons in Laue diffraction geometry, inside a crystal. The effect results from wavefield interference when incident radiation satisfies the Bragg condition. The spatial oscillation period is the pendellösung length. Experimental studies of neutron pendellösung phenomena were carried out by Shull starting in 1968 (Shull 1968).

K. Finkelstein (✉)
CHESS Wilson Laboratory, Cornell University, Ithaca, NY 14853, USA
e-mail: kdf1@cornell.edu

© Springer Nature Switzerland AG 2021   93
G. Jaeger et al. (eds.), *Quantum Arrangements*, Fundamental Theories of Physics 203,
https://doi.org/10.1007/978-3-030-77367-0_4

In this work SO interaction is included in describing neutron dynamical diffraction, and effects on pendellösung oscillations are reported. A resonant enhancement of SO scattering can occur when an external magnetic field applied to the diffracting neutron causes spin precession matched to the pendellösung length. We refer to this effect as Neutron Spin-Pendellösung Resonance (NSPR) (Finkelstein 1987).

Following Horne (Horne et al. 1988) we describe the neutron-crystal potential including nuclear, SO, and external magnetic field interactions, present the corresponding Taupin-Takagi equations for dynamical diffraction, give solutions for the case of a monochromatic beam incident at the Bragg angle, and use the solutions to predict intensity exiting the crystal in the Bragg scattered direction. We then describe an experiment demonstrating this resonant enhancement manifesting as a large change in pendellösung oscillation intensity observed over a small range of applied field. Our measurements yield a precise value for the ratio of SO to nuclear scattering in silicon. This ratio can be used to ascertain the strength of the silicon atomic electric field, and a value for the atomic scattering form factor; this accomplished by a neutron scattering experiment.

Theoretical Considerations

Consider a neutron at position $\mathbf{r}$ in a perfect crystal with three contributions to the interaction potential.

$$V(\mathbf{r}) = V(\mathbf{r})_{\text{nuclear}} + V(\mathbf{r})_{\text{spin - orbit}} + V_{\text{external field}} \tag{4.1}$$

where $V(\mathbf{r})_{\text{nuclear}} = \frac{2\pi \hbar^2}{m} \sum_j b_j \delta(\mathbf{r} - \mathbf{R}_j)$ neutron scattering length is b_j for nucleus at $\mathbf{R}_j$, $V(\mathbf{r})_{\text{SO}} = \frac{\mu e \hbar}{2(mc)^2}\boldsymbol{\sigma} \cdot (\mathbf{E}(\mathbf{r}) \times \mathbf{p})$ with μ the neutron magnetic moment, $\boldsymbol{\sigma}$ the Pauli spin matrix, $\mathbf{E}(\mathbf{r})$ the electric field at $\mathbf{r}$, and $\mathbf{p}$ the momentum operator, and magnetic field potential $V_{\text{external field}} = \frac{-\mu e \hbar}{2mc}\boldsymbol{\sigma} \cdot \mathcal{B}$ with field $\mathcal{B}$ independent of $\mathbf{r}$. Figure 4.1 illustrates the three potentials in the 1-dimensional lattice.

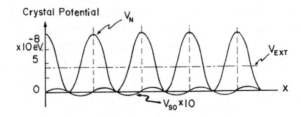

Fig. 4.1 Three potentials in the description of NSPR. The magnitude V_N corresponds to silicon (111) reflection structure factor, V_{SO} depends on atomic electric field and neutron momentum, and V_{EXT} varies with external magnetic field strength. The vertical dashed lines represent crystal lattice planes

Using this potential Mike derived (Horne) the Taupin-Takagi equations describing dynamical diffraction from the time-independent Schrodinger equation (SE). He used a four component neutron wavefunction $\begin{pmatrix} \Psi_+(\mathbf{r}) \\ \Psi_-(\mathbf{r}) \end{pmatrix} = \begin{pmatrix} a_+(\mathbf{r}) \\ a_-(\mathbf{r}) \end{pmatrix} e^{i\mathbf{K_0}\cdot\mathbf{r}} +$ $\begin{pmatrix} b_+(\mathbf{r}) \\ b_-(\mathbf{r}) \end{pmatrix} e^{i(\mathbf{K_0}+\mathbf{G})\cdot\mathbf{r}}$ similar to that in a general introduction to spin-dependent effects on dynamical diffraction (Schlenker and Guigay 2006). Here $\pm$ specify spin along/against the applied field. $\mathbf{K_0}$ is the neutron wavevector, and $\mathbf{G}$ is a reciprocal lattice vector. If wavefield amplitudes vary slowly on the scale of neutron wavelength then terms in SE of order $\nabla^2 a(\mathbf{r})$ can be ignored (Kato 1973) and Fourier expanding the potential in reciprocal lattice vectors $\mathbf{0}$ and $\mathbf{G}$ gives the 2-beam approximation of dynamical diffraction in the plane containing incident beam and $\mathbf{G}$. Let this be the x–z plane with $\mathbf{G}$ along $\hat{\mathbf{x}}$.

Assuming spin quantization axis in the diffraction plane along $\mathcal{B}$, the Taupin-Takagi equation for diffraction in a monoatomic lattice are written in matrix form as:

$$\begin{pmatrix} -i\left(\frac{\partial}{\partial z} + \tan\theta_B \frac{\partial}{\partial x}\right) - \delta & 0 & K\cos\theta & -K\sin\theta \\ 0 & -i\left(\frac{\partial}{\partial z} + \tan\theta_B \frac{\partial}{\partial x}\right) + \delta & K\sin\theta & K\cos\theta \\ K\cos\theta & K\sin\theta & -i\left(\frac{\partial}{\partial z} - \tan\theta_B \frac{\partial}{\partial x}\right) - \delta & 0 \\ -K\sin\theta & K\cos\theta & 0 & -i\left(\frac{\partial}{\partial z} - \tan\theta_B \frac{\partial}{\partial x}\right) + \delta \end{pmatrix} \begin{pmatrix} a_+ \\ a_- \\ b_+ \\ b_- \end{pmatrix} = 0$$

$$(4.2)$$

where $K = \frac{mV(G)}{\hbar^2 K_0 \cos\theta_B} = \frac{\pi}{\Delta}$, with V(G) the nuclear scattering potential Gth Fourier coefficient, Δ is the pendellösung length, $\delta = \frac{m\mu B}{\hbar^2 K_0 \cos\theta_B}$ the magnetic field parameter, θ_B the Bragg angle, and $\tan\theta$ is the ratio of spin–orbit to nuclear scattering amplitudes.

Herb Bernstein (Bernstein) derived general planewave solutions for these equations, but Mike's paper (Horne et al. 1988), and analysis of measurements reported here assume incident neutrons exactly satisfying Bragg's law. In this case the solutions are independent of x and the wavefields are written:

$$\alpha'_+ = e^{i\sigma z}[(K\cos\theta + \delta + \sigma)\begin{pmatrix} 1 \\ 0 \\ -1 \\ 0 \end{pmatrix} + K\sin\theta \begin{pmatrix} 0 \\ 1 \\ 0 \\ 1 \end{pmatrix}]$$

$$\alpha'_- = e^{i\mu z}[(K\cos\theta - \delta + \mu)\begin{pmatrix} 0 \\ 1 \\ 0 \\ -1 \end{pmatrix} - K\sin\theta \begin{pmatrix} 1 \\ 0 \\ 1 \\ 0 \end{pmatrix}]$$

$$\beta'_+ = e^{i\mu z}[(K\cos\theta - \delta + \mu)\begin{pmatrix}1\\0\\1\\0\end{pmatrix} + K\sin\theta\begin{pmatrix}0\\1\\0\\-1\end{pmatrix}]$$

$$\beta'_- = e^{-i\sigma z}[(K\cos\theta + \delta + \sigma)\begin{pmatrix}0\\1\\0\\1\end{pmatrix} - K\sin\theta\begin{pmatrix}1\\0\\-1\\0\end{pmatrix}] \qquad (4.3)$$

The prime indicates x-independent solutions, the column vectors correspond to

Eq. (4.2) by $\begin{pmatrix}a_+\\a_-\\b_+\\b_-\end{pmatrix}$, and $\sigma = +[K^2 + \delta^2 + 2K\delta\cos\theta]^{1/2}$, while $\mu = +[K^2 + \delta^2 - 2K\delta\cos\theta]^{1/2}$.

Theoretical Predictions

In Laue diffraction geometry (reflection planes perpendicular to the entrance surface), wavefields inside the crystal are a linear combination of Eq. (4.3) wavefields matched

to perfect Bragg (PB) spin-up or spin-down incident planewaves $\begin{pmatrix}1\\0\\0\\0\end{pmatrix}$ or $\begin{pmatrix}0\\1\\0\\0\end{pmatrix}$ at

$z = 0$. This gives wavefield amplitudes exiting a crystal of thickness z in the Bragg direction

$$\Psi_+^{PB}(z,\delta) = \frac{1}{2}\left\{\begin{pmatrix}1\\0\end{pmatrix}[\cos(\mu z) - \cos(\sigma z)] - i\left[\begin{pmatrix}A+\delta\\-H\end{pmatrix}\frac{\sin(\sigma z)}{\sigma} + \begin{pmatrix}A-\delta\\-H\end{pmatrix}\frac{\sin(\mu z)}{\mu}\right]\right\}$$

$$\Psi_-^{PB}(z,\delta) = \frac{1}{2}\left\{\begin{pmatrix}0\\1\end{pmatrix}[\cos(\sigma z) - \cos(\mu z)] - i\left[\begin{pmatrix}H\\A+\delta\end{pmatrix}\frac{\sin(\sigma z)}{\sigma} + \begin{pmatrix}H\\A-\delta\end{pmatrix}\frac{\sin(\mu z)}{\mu}\right]\right\} \qquad (4.4)$$

where $H = K\sin\theta$ and $A = K\cos\theta$. These expressions have the same absolute magnitude squared, so Bragg diffracted intensity exiting the crystal is independent of incident polarization and thus unpolarized radiation can be used to study NSPR. Exit intensity (ignoring absorption in the crystal) is

$$I^{PB}(z,\delta) = \frac{1}{2}\left\{1 - \cos(\sigma z)\cos(\mu z) + \frac{(A^2 - \delta^2 + H^2)}{\sigma\mu}\sin(\sigma z)\sin(\mu z)\right\} \qquad (4.5)$$

This expression has several consequences: when H = 0 it is independent of δ and reduces to the usual z-dependent pendellösung intensity, when z = (integer + ½)Δ intensity is zero except very close to δ = A (when Larmor precession and pendellösung lengths are equal), and the magnetic resonance width depends on ratios $\frac{H}{A}$ and $\frac{z}{\Delta}$.

Experiment

NSPR is explored by measuring a sharp variation in Bragg diffracted exit intensity as magnetic field at the neutron is scanned. Baseline intensity, far from resonance, is governed by the underlying pendellösung diffraction oscillations, so before observing NSPR we characterized the pendellösung effect. This requires that phase differences between crystal wavefields be well defined and implies: crystal thickness, wavelength band accepted, and range of in-crystal waves observed at the detector be well defined.

Sample Preparation

Experiments were performed with a perfect silicon crystal oriented for symmetric Laue transmission using the (111) reflecting planes. Crystal thickness was controlled by a polishing method that guarantees front and back faces are parallel; the crystal thickness is t = 5.5524(8)mm. Bragg's law links wavelength bandwidth to accepted beam angle collimation. Two angle limiting slits are used to define bandwidth, one placed on the crystal entrance face (width 0.15 mm), the other (width 1 mm) located 1.355 m upstream, between the sample and graphite monochromator. This slit configuration limits angle acceptance below 3 arcmin, insuring the silicon crystal will be excited over wavelength band $\frac{\Delta\lambda}{\lambda} = 0.57\%$. The critical entrance slit was fabricated from two sheets of cadmium (thickness 0.9 mm) with tapered edges defining the slit opening. Radiation passed by the mono was filtered for high energy harmonics by a thick block of polycrystalline graphite, and transparency of the cadmium was further reduced by secondary oversize boron-plastic slits added over the assembly. To select rays corresponding to near perfect Bragg incidence a duplicate of the entrance slit assembly was attached on the crystal exit face. A reference surface (parallel to the (111) planes) on the side of the crystal was used to align and make parallel the front and back slits. The mean angle of passage of radiation through these slits is zero. This configuration limits uncertainty in phase to 1 part in 300 over 25 Pendellösung oscillation periods typical in these measurements. This uncertainty dominates over other effects and represents the limiting resolution of this system.

Control of Magnetic Field

The precision of magnetic field control is set by three features of the resonance: associated field strength, resonance width, signal size or relative change in neutron intensity. The resonance condition, equivalent to $\mu\, B = V(\mathbf{G})$, is independent of wavelength, and depends only on the nuclear scattering amplitude. Using accepted values for coherent nuclear scattering length and Debye–Waller factor for silicon (111) the nuclear scattering length is $0.41054(8) x 10^{-12}$cm, corresponding to a resonant magnetic field 6279 Gauss.

For Bragg angle 8.7°, $\theta_{SO} \approx \frac{1}{120}$, and crystal thickness 24.5 Δ, NSPR resonance FWHM is expected to be about 4% of the resonance field. To resolve this the magnetic field must be tunable, repeatable, and stable at the 0.5% or about 30 Gauss level. We chose to scan field from below to above the resonance over a range of 12% to include a measurement of baseline intensity. Effects of crystal thickness, the large sacrifice in intensity associated with collimation required to observe Pendellösung fringes, and incident flux at the selected wavelength combine to limit diffracted intensity to about 6 neutrons/minute off resonance, rising to 10 neutrons/minute at the peak. Half hour count time per field point was chosen to limit incident flux variations. This time scale is a compromise between short (1 min) power fluctuations and long (6–12 h) drifts in reactor power. Full field scans taking 12 h were repeated 6–8 times per week.

Temperature Control System

The importance of temperature control was appreciated only after the experiment was underway. NSPR study began by ascertaining, at zero field, the location of pendellösung wavelength fringes by scanning crystal angle over the wavelength band available. A central fringe minimum was chosen to characterize the stability of pendellösung intensity over time. This intensity followed a daily cycle, closely correlated with changes in room temperature. Subsequent experiments, involving heating of the crystal environment, suggested pendellösung intensity is strongly influenced by temperature, both gradients in the space around the crystal, and in time. To overcome effects of temperature instability, the crystal and its plastic pedestal were thermally isolated by wrapping in multiple layers of aluminized Mylar "super-insulation". In addition, the electromagnet coils were isolated to reduce convective heat transfer, and a temperature sensor was mounted on one coil cooling plate and used to control chilled water flow from the Neslab RTE-4 magnet cooler. This arrangement stabilized crystal temperature better than $\pm$ 0.1 °C throughout the 25% change in power dissipation of a typical field scan.

The NSPR Experimental Arrangement

An overview of the experimental arrangement is presented in Fig. 4.2. Thermal neutrons from the MIT reactor pass through a soller slit collimator and are diffracted by a (002) pyrolytic graphite monochromator. The collimator limits horizontal beam divergence to 0.35° and the monochromatic beam has 4% wavelength band at 1A. This beam is then restricted at 16.5 cm by a 1 mm horizontal defining boron-plastic slit, followed by a 42 cm long steel channel (not shown) limiting vertical beam height to 2.5 cm. The channel has 3 mm horizontal downstream opening. As indicated in Fig. 4.2, the crystal, magnet, and support table were enclosed in a styrofoam box with one inch thick walls. Neutrons enter and exit this enclosure through pairs of aluminized Mylar windows separated by the wall thickness. The silicon sample crystal is centered between poles of an electromagnet with field pointing normal to (111) reflecting planes. Magnet and crystal are mounted on the vertical Θ-axis of a triple axis spectrometer. This axis is aligned with the 0.15 mm horizontal defining entrance slit on the front face of the sample. Rotating about the co-aligned 2Θ-axis is a 2.5 m arm supporting a drum shielding the neutron detector. A 70 cm masonite shield limits detector view to the silicon crystal exit slit. Shull measured Pendellösung

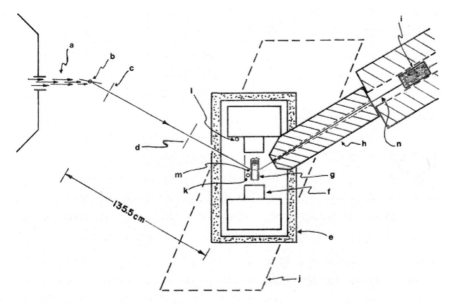

Fig. 4.2 Overview of NSPR experiment (looking down on diffraction plane): **a** thermal neutron beam from reactor, **b** mosaic monochromator crystal rotation axis, **c** 1 mm boron-plastic slit, **d** oversized, 3 mm, beam defining slit, **e** thermal shield around specimen crystal and electromagnet **f**, **g** specimen (silicon (111) reflection), **h** masonite shield passing diffracted beam, **i** neutron detector, **j** spectrometer track, **k** Hall probe, **l** temperature sensor (for magnet cooling water temperature control), **m** specimen rotation axis (aligned with entrance slit), **n** diffracted (signal) beam entering detector

fringes by rotating his sample crystal about the entrance slit as described here, but the incident beam had uniform intensity over a broad wavelength range. Our narrow-band beam was chosen to minimize background, but this limitation proved troublesome in the following respect. The intensity measured in rotating the sample crystal is a convolution of the pendellösung oscillation pattern and (approximately rectangular) wavelength band reflected by the monochromator. Edges of this band are not sharp so edges of the pendellösung wavelength scan become distorted, making assessment of fringe location difficult.

Magnet System

The field between poles of the electromagnet was mapped using a Hall probe with 1 square mm sensing area. In the central section of the one inch gap, over the region of neutron propagation in the silicon crystal, the field was uniform to within 0.04% at the expected resonance field. A Hewlett Packard 6269B power supply, used as a voltage programmable, constant current source provided magnet current. A digitally controlled voltage reference, built for this experiment, produced 30 Gauss field steps with about 6 Gauss precision. After stabilizing cooling water temperature (discussed above), we found precision of magnet current limited field stability during scans. Field strength, monitored by a Hall probe mounted next to the crystal, was found reproducible to within 6 Gauss, and Hall current was stable to 0.0013%/week.

Experimental Results

We measured Bragg diffracted intensity as magnetic field was scanned through the resonance field region. In all, 14 data sets were collected, each representing about 80 h of neutron counting. Six of the 14 were measured at wavelength corresponding to pendellösung minimum, five at an inflection point of intensity, and three at a zero-field maximum. All measurements were made with unpolarized neutrons and no polarization analysis was performed on the diffracted signal. Each data set consists (typically) of six complete magnetic field scans; data at corresponding field points were averaged and standard deviation of the mean was recorded. NSPR theory, corresponding to perfect Bragg incident beam, was used to fit individual data sets by way of a Chi-squared minimization procedure. From this fitting, we obtain a measure of how well theory agrees with our data, and a value for θ_{SO} (the ratio of spin–orbit to nuclear scattering amplitudes).

Representative Data

Figure 4.3 shows pendellösung intensity oscillations (Bragg diffracted exit radiation), at zero applied field, in the wavelength region where NSPR measurements were performed. This information allows us to center the wavelength band at pendellösung phase points where intensity is minimum (crystal thickness $t/\Delta_0 = 25$, silicon (111) reflection, Bragg angle $\approx 8.76°$), at an adjacent intensity half maximum ($t/\Delta_0 = 24.75$), and at intensity maximum position ($t/\Delta_0 = 24.5$). Figures 4.4a, b, c show results of data sets obtained by magnetic field scanning at these pendellösung phase locations. The horizontal axis is the normalized field parameter $\frac{\delta}{K\cos(\theta)}$, the vertical axis is Bragg diffracted intensity normalized to incident flux using a small upstream monitor detector.

Magnet scans were always done in increasing field direction. After each scan magnet current was decreased so that hysteresis did not shift the horizontal axis between scans. Baseline intensity at each wavelength, far from resonance, corresponds to the zero field Bragg diffracted intensity. Many scans were collected and averaged to produce composite scans (data sets) with reasonably small error. The error bars in Fig. 4.4a, b indicate internal error for a sets with six scans each, while Fig. 4.4c represents seven scans. In general, fluctuations in these data were slightly (1–1.5 times) larger than pure statistical error; points collected at extreme ends of the scan show largest fluctuations. This may be due to small changes in crystal temperature when magnet current is far from the central value. In Fig. 4.4a, b the intensity baseline (on either side of resonance) has slight asymmetry. NSPR theory predicts resonance peak shape is very sensitive to pendellösung phase; for example asymmetry visible in Fig. 4.4a can be explained by a crystal angular error from the precise pendellösung maximum of $0.006°$.

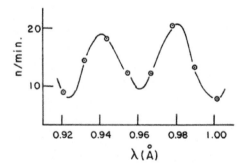

Fig. 4.3 Results of measuring Pendellösung fringe intensity (silicon (111) reflection) as neutron wavelength changes. NSPR measurements were performed around the central minimum where crystal thickness corresponds to 25 Pendellösung lengths

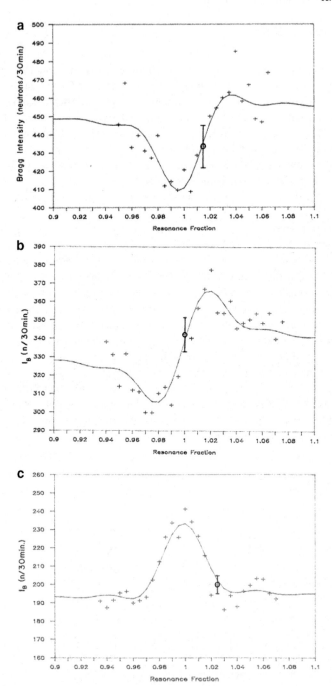

◄**Fig. 4.4** **a** Average of 6 NSPR magnet field scans at wavelength corresponding to zero field pendellösung intensity maximum, or crystal thickness (t) satisfies $t/\Delta_0 = 24.5$. Horizontal axis is normalized field parameter δ/γ. Error bar is standard deviation and solid line is a best fit to theory for perfect Bragg incidence discussed in main text. **b**. Results, averaging 6 NSPR magnet field scans, with crystal angle selecting wavelength corresponding to pendellösung $\frac{1}{2}$ intensity point where $t/\Delta_0 = 24.75$. **c** Results, averaging 7 NSPR magnet field scans, with crystal angle selecting wavelength corresponding to pendellösung intensity minimum, $t/\Delta_0 = 25$

Analysis

Solid lines in Fig. 4.4a, b, c result from a 3-parameter fit to these data, using Eq. (4.5), for each observed wavelength. The 3-parameters are: baseline (zero field Pendellösung minimum) intensity, Pendellösung oscillation amplitude (peak to base-line intensity difference), and θ_{SO}. The fitting was limited in parameter space so that baseline and oscillation intensities are consistent with measurement. In addition to determining θ_{SO} the fit estimates uncertainty for each parameter by calculating Chi-square surface curvature (along each parameter axis) at the minimum.

A first principles value of θ_{SO} comes from the formula (4.6)

$$\theta_{SO} = \frac{\mu\{Z - f(G)\}\frac{e^2}{mc^2}\cot\theta_B}{2b_{silicon}} \tag{4.6}$$

Accounting for differences in θ_B at 3 measured wavelengths, Fig. 4.5 presents $\frac{\theta_{SO}}{\cot\theta_B}$ x 10^{+5} from Chi-squared analysis of 14 data sets, representing 1200 h of field scanning.

The experimental mean and standard error from these measurements is 142.2 ± 2.5 (giving equal weight to each measurement). The measurement implies that nuclear scattering is 108.2 times stronger than spin–orbit scattering at wavelength 0.954 Å. This may be compared to 121, the value based on Eq. (4.6) for the silicon (111) reflection where $Z = 14$, atomic scattering form factor $f(G) = 10.38$, and $b_{silicon} = 0.415 \times 10^{-12}$ cm. The first principles value implies nuclear scattering is 128 time stronger than spin–orbit scattering at this wavelength. *Comparing results one finds measured strength of SO scattering is about 17.5% larger than the first principles calculation.* An explanation for the discrepancy is discussed in the Appendix.

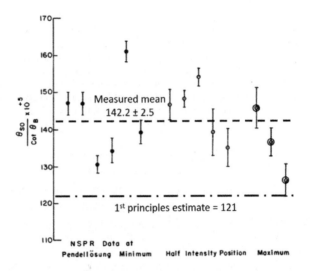

Fig. 4.5 Results of NSPR measurements given as the ratio of spin–orbit to nuclear scattering for silicon (111) reflection. Fourteen data packages, representing about 1200 h of field scanning, are displayed. Vertical axis is normalized to remove wavelength dependence. Three data groups are represented and a weighted average for all packages is shown. The same quantity, derived from first principles calculation, is given for comparison with experiment

Conclusion

We have studied the influence of spin-orbit (SO) scattering on neutron diffraction in silicon by: 1) extending neuron dynamical diffraction to account for 3 interaction potentials, and predicting a resonant effect on Pendellösung intensity we call Neutron Spin-Pendellösung Resonance, 2) the effect is demonstrated by a series of experiments in which unpolarized neutrons diffract through a crystal in an external magnetic field, as it is scanned through the resonance condition, 3) results from these scans are compared to an independent, first principles prediction for the strength of SO scattering in silicon. Our measurements yield a value for SO scattering 17.5% larger than prediction. Finally, we use general, plane wave solutions of the diffraction problem to explore contributions to the signal from a finite range of beam angles not accounted for in the original data analysis.

Acknowledgements The contents of this chapter, and the x-, z-dependent wavefield analysis were done at the Cornell Center for High Energy X-ray Sciences (CHEXS) which is supported by the National Science Foundation under award DMR-1829070. The author thanks Herb Bernstein for reading and helpful advice on this manuscript.

Appendix

The 17.5% discrepancy pointed to in the *Analysis Section* above can be accounted for using general wavefield solutions of the NSPR problem. These planewave solutions are used to match boundary conditions at the crystal entrance surface, when the incident beam does not precisely satisfy Bragg's Law. Equation (A1) lists X-, Z-dependent solutions of Eq. (4.2) derived by Herb Bernstein. Conventions follow *Theoretical Considerations Section*.

$$\alpha''_+ = e^{i(PX + \sqrt{+}Z)} \begin{pmatrix} 1 \\ RB_- \\ R \\ -B_- \end{pmatrix}$$

$$\alpha''_- = e^{i(PX - \sqrt{-}Z)} \begin{pmatrix} RA_- \\ 1 \\ -A_- \\ R \end{pmatrix}$$

$$\beta''_+ = e^{i(PX + \sqrt{-}Z)} \begin{pmatrix} RA_+ \\ 1 \\ -A_+ \\ R \end{pmatrix}$$

$$\beta''_- = e^{i(PX - \sqrt{+}Z)} \begin{pmatrix} 1 \\ RB_+ \\ R \\ -B_+ \end{pmatrix} \tag{A1}$$

$B_- = [\gamma + \delta - \sqrt{+}]/K\mathrm{Sin}\theta_{SO}, \ B_+ = [\gamma + \delta + \sqrt{+}]/K\mathrm{Sin}\theta_{SO},$
$A_- = [-\gamma + \delta - \sqrt{-}]/K\mathrm{Sin}\theta_{SO}, \ A_+ = [-\gamma + \delta + \sqrt{-}]/K\mathrm{Sin}\theta_{SO},$
Also recommend using notation:
$B_\pm = [\gamma + \delta \pm \sqrt{+}]/K\mathrm{Sin}\theta_{SO}, A_\pm = [-\gamma + \delta \pm \sqrt{-}]/K\mathrm{Sin}\theta_{SO}$
$\sqrt{\pm} = \sqrt{[(\delta \pm \gamma)^2 + (K\mathrm{Sin}\theta_{SO})^2]}, \gamma = \sqrt{[(K\mathrm{Cos}\theta_{SO})^2 + (P\mathrm{Tan}\theta_B)^2]},$
$R = Y - \sqrt{(1 + Y^2)}, Y = -(P\mathrm{Tan}\theta_B)/K\mathrm{Cos}\theta_{SO}.$
P is error, from perfect Bragg incidence, of wave vector (X) component parallel to entrance face. All other parameters are defined in the *Theoretical Considerations Section*.

Inner wavefields (α''_-, β''_+) display resonant behavior when $\delta \sim \gamma$, the outer pair do not. Wavefields of Eq. (A1) apply when $\delta \geq \gamma$, i.e. when magnetic field is greater or equal to resonance for perfect Bragg incidence (P = 0). The sign of $\sqrt{-}$ is opposite when $\delta < \gamma$.

Field parameter δ at resonance shifts with γ which in turn depends on P, so off-Bragg incidence radiation excites wavefields satisfying the resonance condition at higher field. The crystal-slit system (*Sample Preparation Section*) limits range of P

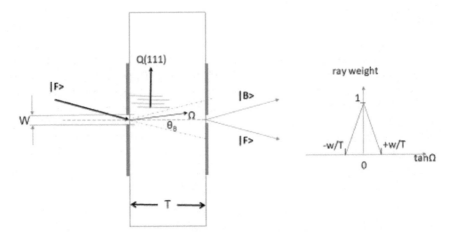

Fig. 4.6 Illustrates beam defining slits of sample crystal, and associated range of ray angles and weight (number of rays) contributing at each internal propagation angle. Incident beam $|F>$ is assumed to be an incoherent sum of plane waves spanning angle range large compared to Darwin width of (111) reflection

contributing signal exiting the crystal, and weights each ray according to propagation angle inside the crystal as illustrated in Fig. 4.6.

Measured strength of SO scattering, *larger* than first principles calculation, can be understood qualitatively by comparing, near resonance in Fig. 4.7, signal versus P/K to that at P/K equals zero. Here we explore the case corresponding to Fig. 4.4c where $T/\Delta_0 = 25$, a zero-field pendellosung intensity minimum. Signal rises away from $P/K = 0$. This *spatial pendellösung* effect increases signal size measured through finite slit openings. Our data analysis used pendellosung peak-to-valley intensity ratio determined by scanning neutron wavelength. NSPR signals are assumed to scale with this difference, but intensity measured at each magnetic field is slightly increased by acceptance of signals away from the $P/K = 0$ minimum. The net result, increased resonant intensity at pendellösung minimum, increased resonant dip at pendellösung maximum ($T/\Delta_0 = 24.5$) are consistent with larger θ_{SO} in our data reduction. The 17.5% discrepancy can be quantitatively accounted for by integrating signal over $|P/K| \leq 0.71$, a range consistent with crystal slit width, yields the signal expected for Eq. (4.5) when $\theta_{SO} \simeq \frac{1}{102}$. Going further with this analyses is beyond the scope of this chapter.

Personal Reflection

The work reported here is a small example of synergy at Cliff Shull's MIT neutron diffraction lab. Mike Horne was a most welcome weekly visitor, who inspired and mentored grad students, as he participated (often motivated) animated discussions, on every conceivable subject (including physics) with visiting scientists, students, and anyone who stopped by the office at NW13. This work evolved from Cliff's

(a)

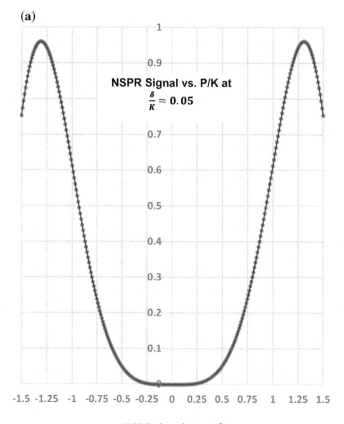

(b)

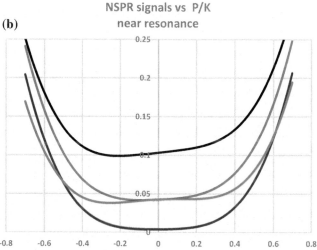

Fig. 4.7 **a** Calculated intensity (vertical axis) exiting crystal in the Bragg diffracted direction, over a range of transverse incident wave vector P/K (horizontal axis). Signal exhibits spatial Pendellösung behavior [5]. Calculation corresponds to Pendellösung minimum $T/\Delta_0 = 25$, when P/K = 0. δ/K = 0.05 i.e. magnetic field 5% of resonance. The region close to central minimum corresponds to range of P/K in our experiments. Signal from adjacent maxima and steeply rising regions are excluded by slits on front and back side of crystal. **b** Calculated NSPR intensity, with increased P/K wave vector resolution, near resonance ($\delta/K = 1$). Signal (points on the U-shape curves) corresponds to δ/K = 0.95 (blue), 0.98 (red), 1.0 (black), 1.02 (green) with Bragg angle 8.7 degrees and $\theta_{SO} \approx \frac{1}{102}$

interest in pushing measurement sensitivity to explore unexpected physics. It is my understanding that he came up with the NSPR concept. Mike worked out the dynamical diffraction description and found the "on-Bragg wavefields" using symmetry and energy conservation. Herb Bernstein used mathematical wizardry to derive general solutions. Meantime, Anton Zeilinger was my early mentor on experimental neutron diffraction methods. Tony Klein, a visitor from Australia, dispensed sage wisdom on the MIT qualifying exams.

Every Tuesday Mike would arrive by 10 AM, inject a shot of insulin, take out a notebook and pen, put up his feet, and get to work. Notebook entries always seemed to contain finished ideas; I cannot remember him using a scratch pad. Mike was a master teacher, patient, able to reduce problems to whatever level the student needed. He excelled at finding interesting explanations, and enjoyed recalling how he used them in lectures at Stonehill College. With everything he accomplished, social distance vanished in Mike's relationships.

Mike taught me most of what I know about dynamical diffraction. Visualization was the key to understanding what went on inside the crystal and what the equations said. He was the most approachable person I met at MIT (after my wife). His wisdom extended way beyond physics. He had a refined palette that analyzed the ingredients in chili I'd bring back for him from MIT pushcart vendors. Mike loved music; he introduced me to SunRa and Robert Cray, played a very cool double bass, and he and Carole made a tape entitled "Our love is here to stay" that accompanied us on our honeymoon...

Editors' note: Mike remained an active member of the Shimony Group at Boston University throughout his career and encouraged the further exploration of his neutron physics work. One of the results of this was the derivation from first principles, in 1989, of assumptions made in Horne et al. (1988), later published as: Jaeger, Shimony (1999) 'An extremum principle for a neutron diffraction experiment' Found Phys 29: 435.

References

Bernstein HJ, Private communication (1986)
Ewald PP (1916) Ann Phys (Leipzig) 49:117; 54:519 (1917)
Fermi E (1954) Nuovo Cim 11:407–411
Finkelstein KD (1987) 'Neutron spin pendellosung resonance' Ph.D. Thesis, Massachusetts Institute of Technology, Cambridge, Mass
Horne MA. Personal notebooks, private communication (1987)
Horne MA, Finkelstein KD, Shull CG, Zeilinger A, Bernstein HJ (1988) Phys B 151:189
Kato N (1973) Naturforsch, 28a:604
Schlenker M, Guigay JP (2006) Paper: international tables of crystallography, vol B. Chapter 5.3, pp 557–569
Schwinger J (1948) Phys Rev 73:407
Shull CG (1963) Phys Rev Lett 10:297
Shull CG (1968) Phys Rev Lett 21:1585

Chapter 5
Mike Horne and the Aharonov-Bohm Effect

Daniel M. Greenberger

Introduction

I first met Mike Horne, together with Anton Zeilinger and Cliff Shull, at a meeting of nuclear physicists at the large French reactor at Grenoble, in 1978. We really hit it off, and this has led to a forty year collaboration between Mike, Anton, and me, which was actively ongoing until Mike's recent untimely death. Researchwise, Mike was like a bulldog, who would not let a topic go until he understood it thoroughly. This means even after writing a definitive paper on a subject, Mike would return to it over and over, and new aspects of the topic would continually emerge. Of course Mike was also a wonderful human being, modest and self-effacing, and I considered him one of my closest friends. One of the worst aspects of my having reached an age into the middle eighties, is this continuing loss of companions who have acted as the steadiest beacons guiding me forward through life. Mike was irreplaceable, and leaves a giant hole in my being.

One of those topics that Mike was continually returning to, was the Aharonov-Bohm (A-B) effect. He felt that people did not appreciate it, and that even though there was no force present acting on the particle, nonetheless the system had to have been prepared, and during this stage, there were forces, and so the effect was not as mysterious as it appeared to be. We will illustrate this by considering one well-known case of the A-B effect. Since I presume that many readers of this paper will be philosophers or historians of science, rather than physicists, I will discuss the topic in more detail than usual.

D. M. Greenberger (✉)
City College of the City University of New York, New York, USA
e-mail: greenbgr@sci.ccny.cuny.edu

© Springer Nature Switzerland AG 2021
G. Jaeger et al. (eds.), *Quantum Arrangements*, Fundamental Theories of Physics 203,
https://doi.org/10.1007/978-3-030-77367-0_5

The Aharonov-Bohm Effect

Using Potentials to Describe Magnetic Fields

In quantum mechanics the treatment of electromagnetic fields occurs more naturally, by using potentials, rather than the fields themselves. The effect of a magnetic field on the system is to change the momentum operator p into $\left(p - \frac{e}{c}A\right)$, where A is the vector potential. Now p as an operator in quantum theory is represented by $p = \frac{\hbar}{i}\frac{\partial}{\partial x}$, so that acting on ψ, it becomes $p\psi \rightarrow \left(p - \frac{e}{c}A\right)\psi$. One might think, looking at this, that one could completely eliminate this effect by writing $\psi = \psi_0(exp\frac{ie}{\hbar c}\int^x A \cdot d\ell)$. Then

$$\psi = \left(p - \frac{e}{c}A\right)\psi_0 e^{(ie/\hbar c)\int A \cdot d\ell}$$

$$= (p\psi_0)e^{(ie/\hbar c)\int A \cdot d\ell} + \psi_0\left(\frac{\hbar}{i}\frac{ie}{\hbar c}Ae^{(ie/\hbar c)\int A \cdot d\ell}\right) - \frac{e}{c}A\psi_0 e^{(ie/\hbar c)\int A \cdot d\ell}$$

$$= (p\psi_0)e^{(ie/\hbar c)\int A \cdot d\ell},$$

so the entire effect of the e-m field would just reduce to a phase factor, which would cancel out when one took matrix elements,

$$\left\langle\psi\left|f\left(x, \left(p - \frac{e}{c}A\right)\right)\right|\psi\right\rangle = \langle(\psi_0|f(x, p)|\psi_0)\rangle.$$

Thus, the effects of the e-m field have disappeared altogether. So an e-m field has no effect at all on the system.

Of course this cannot be correct. So where is the mistake? Actually, it is only partly wrong. If in fact there does exist a function, $\varphi(r)$, such that $\int A \cdot d\ell = (r)$, then the argument will be correct, and the gradient of the integral will exist, and equal A. But the criterion that $\phi(r)$ exists is that $(\nabla \times A) = 0$. But by definition, $\nabla \times A = B$, and so if $\phi(r)$ exists, then $B = 0$. So if $B = 0$, then $\int A \cdot d\ell$ is independent of the path of integration (which is the criterion that all paths lead to the same answer $\phi(r)$). Otherwise $\int_P^x A \cdot d\ell$ is a path-dependent quantity. It depends on the path P and is not a unique $\phi(r)$, and it happens when $B \neq 0$. So one may symbollically write $\psi = \psi_0 exp\frac{ie}{\hbar c}\int_{P(\ell)}^x A \cdot d\ell$, but one must remember that it is path-dependent, as are its matrix elements.

One consequence of this is that if one has a beam that one splits into two beams ψ_1 and ψ_2 which one later recombines (Fig. 5.1a), one may write $\psi_1 = \psi_0 e^{(ie/\hbar c)\int_{P(1)}^x A \cdot d\ell}$ and $\psi_2 = \psi_0 e^{(ie/\hbar c)\int_{P(2)}^x A \cdot d\ell}$ and, then,

$$\psi_2 = \psi_1 e^{(ie/\hbar c)(-\int\limits_{P(1)}^x A \cdot d\ell + \int\limits_{P(2)}^x A \cdot d\ell)}$$

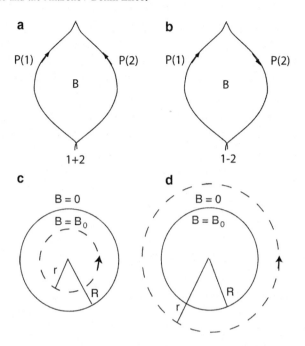

Fig. 5.1 a The sum of the two phase shifts will add up when the two beams recombine. **b** The difference between the two phase shifts is equivalent to an integral that loops the entire area, a closed line integral, equivalent by Stokes theorem to the flux through the area. **c** Inside the loop A increases with r, as the loop encloses only part of the flux. **d** Outside the loop, A decreases, as all the flux is enclosed. This leads to B = 0 outside

$$= \psi_1 e^{(ie/\hbar c)\oint_P^x A\cdot d\ell} = \psi_1 e^{-(ie/\hbar c)\int B\cdot dS},$$

where the integral encloses the entire area between the two beams, (see Fig. 5.1b), and therefore is equal to the flux through the area, by Stokes' theorem, which is a gauge-independent quantity, i.e., it depends only on B, which is path-independent, and no longer on the potential A, which does depend on the path. (Stokes' Theorem says $\oint_{P(\ell)}^x A \cdot d\ell = \oint B \cdot dS =$, the flux through the area enclosed.)

As an example, consider a cylinder with a uniform field B_0 inside, while $B = 0$ outside. (see Fig. 5.1c. The general situation will ultimately be as shown in Fig. 5.6). Here, $B = B_0 \hat{k}$ inside the solenoid, and $B = 0$ outside. (We are looking down on the cylinder, and $\hat{k}$ points up, out of the paper.) It is much easier to analyze this in polar coordinates. Inside the solenoid ($r < R$, see Fig. 5.1c), $\vec{A_0} = A_\varphi \hat{\varphi}$. From Stokes' Theorem,

$$A_\varphi(2\pi r) = B_0\left(\pi r^2\right), \quad A_\varphi = \frac{1}{2}B_0 r, \ \vec{A_0} = \frac{1}{2}B_0 r \hat{\varphi}.$$

Outside the solenoid ($r > R$, Fig. 5.1d), $\overrightarrow{A_0} = A_\varphi \hat{\varphi}$. Again from Stokes' Theorem,

$$A_\varphi(2\pi r) = B_0 \pi r^2, \quad A_\varphi = \frac{\frac{1}{2} B_0 R^2}{r}, \quad \overrightarrow{A_o} = \frac{\frac{1}{2} B_0 R^2}{r} \hat{\varphi}.$$

So outside, $A_o \neq 0$, even though $B = 0$. (In Cartesian coordinates, $\overrightarrow{A_0} = \frac{R^2}{2} \frac{B_0 \times r}{r^2}$. So that, $A_{0,x} = -\frac{B_0 R^2 y}{2(x^2+y^2)}$, $A_{0,y} = \frac{B_0 R^2 x}{2(x^2+y^2)}$, and $B = \nabla \times A_0 = 0$.) But even though $B = 0$ outside, $\int A_0 \cdot d\ell$ nonetheless gives the total flux of the field inside the solenoid, $\left(\int A_0 \cdot d\ell = \frac{R^2 B_0}{2} \frac{\hat{\varphi}}{r} \cdot (2\pi r \hat{\varphi}) = B_0 R^2 \right)$. The constant flux does not contribute to a B field outside.

The Direct Use of Magnetic Fields

Besides the potentials, one can use the actual fields to reveal the behavior of the particles. Classically, when a charged particle moves through a magnetic field, it bends in a circle. In moving through a distance the angle by which it is deflected is (see Figs. 5.2, 5.3 and 5.4)

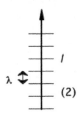

Fig. 5.2 With no field present, the wave fronts travel in straight lines. A distance λ is equivalent to a phase difference of 2π

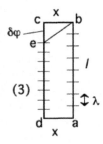

Fig. 5.3 With a field present, the flux enclosed by the trapezoid *deba* will induce a gradient in λ so that the beam will bend, according to Faraday's law

Fig. 5.4 The gradient in the phase will produce the classical bending of the beam

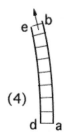

(4)

$$\alpha = \frac{\delta p}{p} = \frac{F\delta t}{p} = \frac{\frac{e}{c}vB\delta t}{p} = \frac{evB}{cp}\frac{\ell}{v} = \frac{eB\ell}{cp},$$

where ℓ is the distance through which the particle would move with no field present, and the Lorentz force is $F = \frac{ev \times B}{c}$, with B in the direction into the paper.

This will be $\delta\ell = \delta\varphi\frac{\lambda}{2\pi} = \frac{e}{\hbar c}Bx\ell\frac{\lambda}{2\pi} = \frac{eBx\ell}{c}\frac{\lambda}{h} = \frac{eBx\ell}{cp}$ (see Fig. 5.3). This corresponds to a bending through the angle β given by $\beta = \frac{\delta\ell}{x} = \frac{eB\ell}{cp} = \alpha$ (see Fig. 5.4) so the resulting deflection of the direction of the beam agrees with the classical one, and this is the correct way to include the magnetic field in quantum calculations, $\psi_0 \rightarrow \psi_0 e^{\frac{ie}{\hbar}\int B \cdot dS} = \psi$.

The Aharonov-Bohm Effect

Because of the non-zero vector potential outside the solenoid, the potential causes a phase shift in the wave function which leads to an amazing effect having no classical counterpart, the Aharonov-Bohm effect. (The original A-B paper was Aharonov and Bohm 1959; a fairly complete guide to the early work, experimental and theoretical, on the A-B effect is the book Peshkin and Tonomura 1989). If a beam of electrons is split by a beam splitter so that half passes on either side of an infinite solenoid and a detector picks up the signal from the interference pattern when they combine, the interference pattern will show an extra interference term due to the presence of the solenoid, $\psi_1 + \psi_2 \rightarrow \psi_1 + \psi_2 e^{\frac{ie}{\hbar c}BS}$, where S is the area of the solenoid (see Figs. 5.5 and 5.6).

The problem is that there is no magnetic field outside the solenoid. The field is completely confined to the region inside the solenoid. So, although neither electron beam is *ever* in a magnetic field, there will be a phase shift between them due to the presence of the solenoid. This is a real effect, and it has been convincingly experimentally demonstrated to take place.

Classically, if there is no field present, there can be no force on the electron beams and therefore no effect. Quantum mechanicallly, there is also no force, and the envelope of each of the wave packets for ψ_1 and ψ_2 will not shift, but within that envelope the phase changes, and so the interference pattern will change.

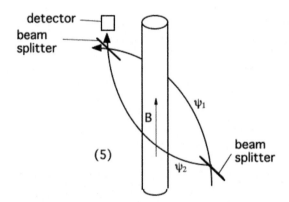

Fig. 5.5 A split beam passing around opposite sides of the solenoid encloses a flux, which produces a phase difference between the beams, which produces interference when they recombine

Fig. 5.6 A top view of Fig. 5.5

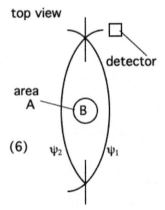

Where does such a quantum effect come from? The answer has to do with the meaning of the wave function. The wave function is the solution corresponding to a particular Hamiltonian operator, and it represents the result of any experiment that can be performed anywhere in space on that system (which is represented by that Hamiltonian). Specifically, it contains implicitly the non-local knowledge generated by the sum of all such experiments. (This argument was first given in Greenberger 1981.)

In our case, we might have performed any of several experiments. Imagine a small hole drilled in the solenoid that allows us to perform the following experiments (see Fig. 5.7). We could have performed our experiment with the beams I and III, Beam I feels no force but beam III passes through the magnetic field and is deflected. So there is a real force on beam III which shows up in the interference pattern through the phase $\psi_I = \psi_{III} e^{\frac{ie}{\hbar c} \oint B \cdot dS_1}$—we are assuming for convenience that the beams contain particles with a + charge.

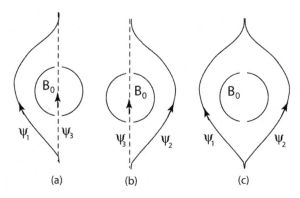

Fig. 5.7 a The experiment can be done between beams I and III. In this case, there is a real field, and clearly there is a phase shift between them. **b** The experiment can also be done between beams III and II. Again there is a real field, which produces the same phase shift between III and II. **c** If the experiment is done between beams I and II, the phase shift between them will be double that in each case above, even though neither beam ever sees a field. The quantum beams keep track of this

Now we could also do an experiment with beams III and II. Again beam III is deflected, corresponding to a real field, but it is deflected away from beam II so the two experiments are not symmetrical. In this case, the interference is caused by the phase $\psi_{III} = \psi_{II}e^{\frac{ie}{\hbar c}\oint B \cdot dS_2}$.

Again, we could do a third experiment with beams I and II, neither of which feels the magnetic field but which are related to each other by $\psi_I = \psi_{III}e^{\frac{ie}{\hbar c}\oint B \cdot dS_1} = \left(\psi_{II}e^{\frac{ie}{\hbar c}\oint B \cdot dS_2}\right)e^{\frac{ie}{\hbar c}\oint B \cdot dS_1} = \psi_{II}e^{\frac{ie}{\hbar c}\oint B \cdot dS_{1+2}}$. So the system keeps track of the phases inherent along all of the paths between the two beams and lets us know that any experiment between beams I and II must be consistent with everything that goes on in between, even though neither beam itself ever sees a magnetic field.

The moral, as we said, is that the wave function represents the solution to the Hamiltonian, and so keeps track of all experiments consistent with the Hamiltonian. It is your choice to make up a wave-packet that only uses waves that don't ever contact the magnetic field, and then you would never know that the magnetic field is there. And this will be true even if you split and recombine the beam. But if the split crosses an area of the space where a magnetic field is present, the phases will automatically take care of that, because you might have crossed the magnetic field in your experiment, and the results must be consistent.

After all, when one chooses to do an experiment with beam II, one makes up a wave packet out of plane waves that sample all of space. There happens to be no field present at the wave packet, so that one could easily be unaware that these waves had been built up of waves which had a phase shift present at the solenoid, representing a real magnetic field there. Had one chosen to continuously move the wave packet from I to II, it would have sampled the solenoid and picked up a phase shift. And once it passed the solenoid the phase shift built in wouldn't just go away. It would add up as a constant phase so that if one made up a wave packet on the other side of the solenoid the same constant $e^{\frac{ie}{\hbar c} \oint B \cdot dS_{1+2}}$ would multiply all components of the wave packet and one again wouldn't notice it. But for two wave packets formed on opposite sides of the solenoid this effect of the presence of the solenoid is there, and it will show up.

How different this is from classical physics, where only local fields that affect the beam can have any influence! Some people choose to say that in quantum theory, unlike classical theory, the electromagnetic potentials, in this case the vector potential, A (where $B = \nabla \times A$), would have a dynamical effect. (A doesn't vanish outside the solenoid even though B does.) But the effect is really deeper than that. Wu and Yang (1975) consider the line integral $e^{\frac{ie}{\hbar c} \int A \cdot d\ell}$ to be the physically meaningful quantity. We would like to point out that the A-B effect reflects the non-local book-keeping of the wave function, which samples all of space. This is a truly non-local effect of quantum mechanics. It is a different type of non-locality from that which shows up in Bell theorem type experiments, where "non-locality" is debatable.

One thing that always interested Mike Horne was that if you started the experiment without any magnetic field present and then built up the magnetic field inside the solenoid, Faraday's law would tell you classically that, while the field was building up, there would be an induced electric field around the solenoid which would be in opposite directions on each side of the solenoid, and the work done by the electric field would exactly add up to the phase shift of the A-B effect so the effect looks classical. But, classically, there is no wave function or phase shift so there is no mechanism for the system to remember that you built up the magnetic field to produce the phase difference on opposite sides of the solenoid and, when the build-up is done, the beams do not enter any field regions and the beams are not deflected.

Mike also went beyond this (Horne 2012) A paper showing some of his concerns in this direction can be found in Horne (2012).

References

Aharonov Y, Bohm D (1959) Significance of the electromagnetic potentials in the quantum theory. Phys Rev 115(3):485
Greenberger DM (1981) The reality and significance of the Aharonov-Bohm effect. Phys Rev D 23:1460
Horne MA (2012) Found Phys 42:140
Peshkin M, Tonomura A (1989) The Aharonov-Bohm effect, Lecture notes in physics, #340. Springer, Berlin
Wu TT, Yang CN (1975) Concept of non-integrable phase factors and global formulation of gauge fields. Phys Rev D 12(12):3845

Chapter 6
The Essence of Entanglement

Časlav Brukner, Marek Żukowski, and Anton Zeilinger

On Mike

By Anton

John Bell and Bernard d'Espagnat organized in 1976 (18–23 April) "Thinkshops in Physics: Experimental Quantum Mechanics" in Erice, Sicily. With the neutron interferometer, we had just done the so-called 4π spin rotation experiment (Rauch et al. 1975) where we confirmed the fundamental prediction of quantum theory that a half-integer spinor would obtain a phase vector of -1 upon a full rotation through 2π. The thinkshop was focused on quantum entanglement as triggered by the recent observation of a Bell-type inequality violation by Freedman and Clauser (1972). When I met Mike there, we immediately became friends, realizing that we have very similar views on the foundations of quantum mechanics, of the challenges ahead, and of the necessity of finding novel experiments confirming the counter-intuitive implications. I will never forget Mike's presentation at that meeting. Very modestly he said that he has nothing to new to add to the talks of earlier speakers. Basically he

---

Dedicated to the memory of Mike Horne.

Č. Brukner (✉) · A. Zeilinger
Institute for Quantum Optics and Quantum Information (IQOQI-Vienna) Austrian Academy of Sciences & Group for Quantum Optics, Quantum Nanophysics and Quantum Information, University of Vienna & Vienna Center for Quantum Science and Technology (VCQ), A-1090 Vienna, Austria
e-mail: caslav.brukner@univie.ac.at

A. Zeilinger
e-mail: anton.zeilinger@univie.ac.at

M. Żukowski
International Centre for Theory of Quantum Technologies (ICTQT), University of Gdansk, PL-80-308 Gdansk, Poland
e-mail: marek.zukowski@univie.ac.at

© Springer Nature Switzerland AG 2021
G. Jaeger et al. (eds.), *Quantum Arrangements*, Fundamental Theories of Physics 203,
https://doi.org/10.1007/978-3-030-77367-0_6

said, in his famous relaxed way, that we do not understand what the stuff emerging from a source of entangled pairs is made of.

In 1977, I joined the group of Cliff Shull—who in 1994 would receive the Nobel Prize in Physics—in the Neutron Deflection Laboratory at M.I.T. There, to my great joy, was Mike on a sabbatical year from Stonehill College. The following years, we worked very closely together, exclusively on neutron interferometry. I was the experimentalist and Mike Horne did much of the theory. His theoretical approach was signified by taking extreme care in two ways. On the one hand, he was not satisfied by having the results of a calculation. He insisted to fully understand all details of what is going on and the conceptual implications. Also, very often, he derived his results in more than one even conceptually different ways. For example, to analyze the effect of gravity on a massive particle interferometer, he developed at least four different approaches which gave beautiful insights into the role of acceleration and of gravity on quantum phases. To my knowledge, this work is still unpublished.

Mike was a scientist, only interested in physics and not at all on promoting himself. I would like to tell a small story. Sometime around 1980, we went to a physics colloquium by Anthony French at M.I.T., who presented the history and the concepts in the classic Fizeau experiment for photons, which demonstrates the dragging of reference frames by moving matter in an interferometer for light. We immediately started to work on the matter wave analog. While we worked on it, we discovered that Asher Peres was also interested in the topic. We had some correspondence with Asher. When he published his paper, Mike said that he was happy not having to publish his analysis anymore!

After my extended stays at M.I.T. in 1977–1978 and 1981–1983, I used to come back to M.I.T. for many summers, where we regularly worked together. One day in the summer of 1984, when I came to the lab, Mike asked me: "Do you want to go to Finland?" My answer was: "Sure, anytime, but why?" Smiling, Mike showed me the announcement of the "Symposium on the Foundations of Modern Physics: 50 Years of the Einstein-Podolsky-Rosen Gedankenexperiment" which was announced for 1985 in Joensuu, Finland. So we said: "Okay, if we want to go, we have to invent a connection between matter wave interference and Einstein-Podolsky-Rosen". In long discussions, we realized that there was a strong mathematical analog between an interferometer and an electron spin. Matter wave interferometers can be seen as two-state devices where the interfering particle follows two different paths. These two paths are analogous to the two states of a spin-$\frac{1}{2}$ system or to the two orthogonal polarization states of a photon. So we clearly had the essence of a qubit in front of us—without naming it that way. The result of this work was the concept of two–particle interferometry as presented at that meeting in Joensuu, Finland (Horne and Zeilinger 1985) (Fig. 6.1).

As relaxed Mike's personality was when it came to scientific priorities, as stubborn he could be when it came to issues he considered to be scientifically very relevant. As an example I still remember the intensive, long, and sometimes quite loud discussions he, Danny Greenberger and myself had in my house in Tirol in 1993, when we prepared our small review note in Physics Today on "Multiparticle Interferometry and the Superposition Principle" (Greenberger et al. 1993). Danny,

Fig. 6.1 Mike Horne (right) in discussion with Danny Greenberger (center) and Anton Zeilinger (left)

Mike and I certainly shared the same view of the fundamental issues, but each of us could be quite persistent when it came to specific formulations. The collaboration with Mike and Danny continued with many visits, either of us crossing the Atlantic. I vividly remember long and deep discussions at Carol and Mike's kitchen table in Boston and in my family vacation place on Traunsee Lake in Austria.

This paper presents some results on the fundamental role of information in understanding entanglement. I hope that it can be seen as a step towards understanding the "stuff" emitted by an entanglement source, as expressed by Mike in 1976 in Erice.

By Marek

I first knew Mike from the literature, the famous CHSH and CH papers. His Joensuu conference paper with Anton was the trampoline which catapulted me into studies of entanglement. I was passively interested in the Bell Theorem and related stuff. Suddenly it turned out that together with Jarek Pykacz, we were able to suggest a feasible realization of the EPR interference gedankenexperiment suggested in the paper. This resulted in a change in the direction of my research. At another conference in Finland (Joensuu 1990) I was jolted by an intellectual shock: the Greenberger-Horne-Zeilinger paradox. So many things were to be done in entanglement! All that ended up with Anton offering me a visiting professorship in Innsbruck. A series of visits started in 1991. Of course, soon Mike was also there. His one sentence remark in June 1992 in Anton's flat put us on the right course concerning our understanding of the interference of independent photons, which, after months, led to the entanglement swapping paper. We were blind, he could see.

On the same day I had a sensorial inconsistency experience. I entered a room where Mike and Anton were listening to Louis Armstrong recordings. Recordings from the pre hi-fi era, low quality of sound, but somehow the break-machine sound was like that from a top-end hi-fi set ... Well, you might have guessed it was Mike's precision jazz drumming. Later on I learned that he was the President of the Boston Jazz Society. Another jazz story. A few years later in a Heuriger, I was grumbling that there is no CD re-issue of *The Modern Jazz Quartet* masterpiece "The Third Stream Music". He sighed, and said "well, next week Milt Jackson will be with us, so I will tell him about that". Thanks to Mike, I was one handshake away from Milt Jackson, the genius of vibraphone!

Thanks to years of collaboration, which were in reality a very short periods during which we overlapped in Anton's institutes, I learned how to reduce and simplify a problem in physics to such an extent that it is almost trivialized, and the solution is transparent. He introduced me to punctuation in American English, showing me how much it can express, and how subtle it a can be (in Polish punctuation follows strict rules). Finally, I shall always remember his "okey-dokey" ....

By Časlav

One of the things that define my memory of Mike is his utterly unpretentious humility. The CHSH Inequality which was named after him and his collaborators has had a lasting impact on quantum physics, but he was distinctly unaffected by the fame and honor that came with that achievement. Rather, he seemed to be a bit vexed by the amount of time the attention he received because it took away from the thing he was really interested in: physics. His driving force was his curiosity about fundamental quantum mechanics, and he seemed to derive almost as much enjoyment from others' insights as from his own findings. Some of his calculations on the concept of multi-particle interference visibility were never published. As soon as he had explored and understood a phenomenon, it was no longer interesting to him. The work involved in the publication process was simply a tedious task he undertook to multiply knowledge by sharing it.

The second thing I remember most about Mike was his ability to develop the mathematically most uncomplicated model of a phenomenon possible, with an elegant austerity that expresses its essence.

Introduction

In their seminal paper (Einstein et al. 1935) Einstein, Podolsky and Rosen (EPR) consider quantum systems consisting of two particles such that while neither position nor momentum of either particle is well defined, both the sum of their positions and the difference of their momenta are both precisely defined. It then follows that measurement of either position or momentum performed on, say, particle 1, immediately implies for particle 2 a precise position or momentum respectively, even when the two particles are separated by arbitrary distances without any actual interaction between them.

Motivated by EPR, Schrödinger in his paper entitled "The present situation in quantum mechanics" (Schrödinger 1935a) wrote succinctly "Maximal knowledge of a total system does not necessarily include total knowledge of all its parts, not even when these are fully separated from each other and at the moment are not influencing each other at all". Schrödinger (Schrödinger 1935a) used the German term "Verschränkung", and he himself introduced the English translation "entanglement" (Schrödinger 1935b). So he coined the expression "entanglement of our knowledge" to describe this situation. A central goal of our work is to take Schrödinger's position as a starting point for a quantitative information-theoretic definition of entanglement. As Schrödinger, and in his works Mike Horne, we restrict our consideration to *pure* states (with some results being valid also for mixed states). Carefully reading his sentence, one may identify three independent ideas on which Schrödinger builds his notion of entanglement.

(a) First, our knowledge, or total information, of a system is bound.
(b) Second, the total information of a composite system is not necessarily fully contained in its individual constituents.
(c) Third, these statements are independent of the relative space-time arrangements of the individual observations on the constituents of the composite system.

Furthermore, we underline that Schrödinger talks about quantum states representing our expectation catalogs and carefully avoids the notion of system properties.

In 1964, John Bell (Bell 1964) obtained certain bounds (Bell inequalities) on combinations of statistical correlations for measurements on two-particle systems if these correlations are understood within a realistic picture based on local hidden properties of each individual particle. In a realistic picture the measurement results are determined by properties the particles carry prior to and independent of observation. In a local picture the results obtained at one location are independent of any measurements or actions performed at space-like separation. Bell then showed that quantum mechanics predicts a violation of these constraints for certain statistical predictions for two-particle systems. A more striking conflict between quantum mechanical and local realistic predictions even for perfect correlations was discovered for three and more particles (Greenberger et al. 1989, 1990, 1990), resulting in an direct invalidation of the EPR concepts. By now, a number of experiments (Freedman and Clauser 1972; Aspect et al. 1981; Pan et al. 2000) have confirmed the quantum mechanical predictions even if the individual particles are truly space-like separated (Weihs et al. 1998). However, two important questions remain: (1) "What are the most general constraints on correlations imposed by local realism?" and (2) "Which quantum states violate these constraints?". The latter has been solved in general only in the case of two particles in pure states (Gisin 1991; Gisin and Peres 1992) and for two-qubit mixed states (Horodecki et al. 1995). Only recently bounds for a local realistic description of higher-dimensional systems were found in some simple cases (Kaszlikowski et al. 2000).

Our paper will now develop the ideas presented above in various quantitative ways. Following first Schrödinger's ideas, we will define quantum entanglement as a feature of a composite system to have more information contained in correlations than any classical mixture of its individual constituents could ever have. The essence of classical correlations is that there, the joint properties can be reduced to correlations between properties of the individual constituents. This is an operational definition because information is always defined through observation of measurement results.

In parallel, following Bell's ideas, we will obtain a single general Bell inequality that summarizes all possible local realistic constraints on the correlations for a multi-qubit system, where two dichotomic observables are measured on each individual qubit. This enables us to introduce another operational definition of entangled states as those which violate that general Bell inequality in a direct measurement.[1] Finally, we show that the two operational definitions are equivalent in the two-qubit case and that in general, our informational definition of entanglement provides a necessary and rather stringent condition for violating the general Bell inequality. We find this intriguing because the two operational approaches are based on completely different concepts. To us, this further supports the view that information is the most fundamental concept in quantum physics (Zeilinger 1999).

Finiteness of Information

A central point in our discussion will be the different ways how information can be distributed within a composite system. We will introduce our notion of information and offer our considerations of how much information a system can represent.

Any physical description of a physical system is a set of propositions together with their truth values—true or false. Then, any proposition we might assign to a quantum system, which is always based on observation of properties of the classical apparatus used, represents our knowledge, i.e., information, of a system gained through observation. To illustrate this, consider the state $|\psi\rangle = |z+\rangle$ of a spin-$\frac{1}{2}$ particle with spin

---

[1] What is often referred to as an entangled state is a state that cannot be represented as a classical mixture of product states (non-separable state). Here, since our aim is to relate Schrödinger's notion of entanglement to that of Bell's inequality, and we limit ourselves to pure states, we will use the definition as given in the text above. However, it is well known that there are cases of "hidden nonlocality" where a quantum state initially does not directly violate a Bell inequality, but after local operations together with classical communication such a violation might occur (Popescu 1995).

up along the z-axis, which is an eigenstate of the operator σ_z with the eigenvalue +1. This simply means that the quantum system described by the state $|\psi\rangle$ will be found with certainty to have spin +1 if it is measured along the z-axis. Thus the information content in that state is represented by the truth value of the proposition: "The spin along the z-axis is up." This is one bit of information. It is clear that both truth values must be possible. Only then can the observation of the system result in a gain of information. In agreement with Schrödinger's idea (a), the most simple system just represents one bit of information. This is what we mean when we talk about a system carrying information. To us, a system is a construct based on information.

If we assume that one bit is the only information the most simple quantum system can carry, and that this is defined with respect to a certain measurement, then other measurements must contain an element of irreducible randomness. Otherwise the system would carry more information in conflict with (a) for a system representing one bit of information only. This means that there are other measurement directions for which the experimental outcome is completely random. Specifically, for a measurement along any direction **n** in the x-y plane, the proposition "The spin along the **n** axis is up" is completely indefinite, i.e., we have absolutely no knowledge which particular outcome "spin up" or "spin down" will be observed in an individual experimental trial. The two propositions about the spin along z and the spin along **n** are propositions with a property of mutual exclusiveness. This is quantum complementarity: the complete knowledge of the truth value of one of the propositions implies maximal uncertainty about the truth values of the other.

How much information is carried by a system with respect to a specific set of mutually complementary propositions? We suggest that it is natural to assume that the information contained in a set of mutually complementary propositions is the sum over the measures of information of that particular set's individual members (Bohr 1958). Specifically, to obtain the total information carried by a quantum system, one summarizes over all individual measures of information for a complete set of mutually complementary measurements, as shown by Brukner and Zeilinger (1999). There, it was shown that the total information carried by the composite system consisting of N qubits in a pure state is N bits of information.

Here, we are interested in the various ways how information can be distributed within a composite system. In particular, we will consider that part of the total information of the system which is contained in correlations, or joint properties of its constituents. This is also the reason why we do not consider complete sets of mutually complementary propositions for the composite system but only a subset of them concerning joint properties of its constituents.

Information Contained in Correlations

Correlations between quantum systems have assumed a very central role in the discussions of the foundations of quantum mechanics. We will now investigate how much information can be contained in such correlations in order to give an information-theoretic criterion of quantum entanglement. It is our final goal to compare that criterion with the one given by Bell-type inequalities (Clauser et al. 1969; Mermin 1990; Belinskii and Klyshko 1993), which involve for each observer a pair of possible measurement settings.

For clarity of presentation, we first investigate the case of a two-qubit system carrying therefore N = 2 bits of information, i.e. representing the truth value of two propositions. The information contained in 2 propositions can be distributed over the 2 qubits in various ways.

Consider first a product state, e.g. $|\psi\rangle = |+x\rangle_1 |-x\rangle_2$. Here, the state $|\psi\rangle$ represents the two-bit combination true-false of the truth values of the propositions about the spin of each particle along the x-axis: (1) "The spin of particle 1 is up along x" and (2) "The spin of particle 2 is up along x". Instead of the second proposition describing the spin of particle 2, we could alternatively choose a proposition which describes the result of a joint observation (3): "The two spins are the same along x." Then, the state $|\psi\rangle$ represents the two-bit combination true-false of the truth values of the propositions (1) and (3). Note that this is also present in classical composite systems.

Evidently, for pure product states, at most one proposition with definite truth-value can be made about joint properties because one proposition has to be used to define a property of one of the two subsystems. In other words, 1 bit of information defines the correlations

$$I_{corr}^{prod} = 1. \tag{6.1}$$

In our example where $|\psi\rangle = |+x\rangle_1 |-x\rangle_2$ the correlations are fully represented by the correlations between x-measurements on both sides, therefore $I_{corr}^{prod} = 1 = I_{xx} = 1$. The specific measure of information we use is specified below.

Obviously, the choice of directions within the planes of measurement on the two sides is arbitrary. However, we will choose the pair of measurement planes for which the information in the correlations is maximal. We will consistently refer to both such planes as x-y. We allow the observers to choose any pair of complementary (orthogonal on the Bloch sphere) directions in the planes x-y. Therefore, in concurrence with Brukner and Zeilinger (2001), we request that the total information contained in the correlations must be invariant upon this choice, as long as one stays within the chosen planes. This invariance property by itself already defines a specific measure of information and rules out Shannon's measure (Brukner and Zeilinger 2001). Following our arguments given above, we now define the information contained in the correlations as the sum over the individual measures of information carried in a complete set of mutually complementary observations within the pair of (optimal) x-y planes. Therefore the information contained in the correlations is quantified by the sum

$$I_{corr} = I_{xx} + I_{xy} + I_{yx} + I_{yy} \tag{6.2}$$

of the partial measures of information contained in the set of complementary observations within the x-y-plane. These observations are mutually complementary for *product states* and the set is complete as there exists no further complementary observation within the chosen planes. By this, we mean that for any product state, complete knowledge contained in one of the observations in Eq. (6.2) excludes any knowledge content in the other three observations.

Consider now a maximally entangled Bell state, e.g.

$$|\phi^-\rangle = \frac{1}{\sqrt{2}}(|+x\rangle_1|-x\rangle_2 + |-x\rangle_1|+x\rangle_2)$$

$$= \frac{1}{\sqrt{2}}(|+y\rangle_1|+y\rangle_2 + |-y\rangle_1|-y\rangle_2). \qquad (6.3)$$

The two propositions expressed here in the two forms of the state (Eq. 6.3) are statements about results of joint observations (Zeilinger 1997), namely (1') "The two spins are equal along x" and (2') "The two spins are equal along y". Now the state represents the two-bit combination false-true of these propositions about correlations. Note that here, the 2 bits of information are all carried by the 2 qubits in a joint way, with no individual qubit carrying any information on its own. In other words, as the two available bits of information are already exhausted in defining joint properties, no further possibility exists to also encode information in individuals. Therefore

$$I_{corr}^{Bell} = 2. \qquad (6.4)$$

Note that in our example $I_{xx} = I_{yy} = 1$ and $I_{xy} = I_{yx} = 0$. Also, note that the truth value for another proposition, namely, "The two spins are equal along z", must follow immediately from the truth values of propositions (1') and (2'), as only 2 bits of information are available. Interestingly, this is also in concurrence with the formalism of quantum mechanics, as the joint eigenstate of $\sigma_x^1\sigma_x^2$ and of $\sigma_y^1\sigma_y^2$ is also the eigenstate of $\sigma_z^1\sigma_z^2 = -(\sigma_x^1\sigma_x^2)(\sigma_y^1\sigma_y^2)$.

In contrast to product states, we suggest entanglement of two qubits to be defined in general so that *more than one bit* (of the two available ones) is used to define joint properties, i.e.

$$I_{corr}^{entgld} > 1 \qquad (6.5)$$

for at least one choice of the local x and y directions for the two qubits. This is in agreement with Schrödinger's idea (b). Equivalently, we suggest to define the two qubits as classically composed (non-entangled) if less than or equal to one bit of information is used to define correlations, i.e.

$$I_{corr}^{nonent} \leq 1 \qquad (6.6)$$

for all possible choices of the local x and y directions for two qubits. As we will show below, the independent information-theoretic definition of entanglement (Eq. 6.5) will turn out to be equivalent to a necessary and sufficient condition for a violation of a Bell-CHSH inequality for two-qubits.

In the generalization to more than two qubits we consider, without loss of generality, a product state $|\psi\rangle = |+x_1\rangle|+x_2\rangle \ldots |+x_N\rangle$ of N qubits. Here, x_j denotes a spatial direction in a local coordinate system of observer j. Then, only one proposition with definite truth-value can be made about the correlations in the N qubits, namely the proposition (*): "The product of spin of particle 1 along x_1, spin of particle 2 along x_2, ...and spin of particle N along x_N is +1". This means that for a product state, again at most one bit of information can be found in N-qubit correlations (i.e. there is a just one set of local settings for which all qubits give the same results with probability 1). We therefore suggest to define N qubits as classically composed if not more than one bit of information is used to define correlations, i.e.

$$I_{corr}^{nonent} = \sum_{\substack{x_1,\ldots,x_N \\ =x,y}} I_{x_1,\ldots,x_N} \leq 1 \tag{6.7}$$

for all possible choices of local x and y directions for N qubits. Here, the sum is over the measures of information over a set of propositions of the type given above where $x_1, \ldots, x_N \in \{x, y\}$ and which therefore are mutually complementary.[2]

We would like to make a very general comment. The quantitative condition (Eq. 6.7), while certainly correct for the situations discussed here, might have to be modified in order to apply to more complicated cases like entanglement between many qubits when the measurements are not restricted to one plane or entanglement between systems defined in Hilbert spaces of higher dimensions, so-called qunits. Another interesting case can arise when considering in detail all possible sets of correlations between all possible sets of subsystems. For example, for 3-qubit systems, we have one correlation between all three individual qubits, we have three correlations between two individual qubits and we have another three correlations between one qubit and the other two. This results in a large number of conditions of the type of Eq. (6.7) which are not independent from each other in general. Yet, we stress that it is to be expected that the general ideas laid out here will still be applicable. A most important guideline for quantitative conditions is that the information carried by the correlations between subsystems exceeds the limit given by the information carried by the subsystems themselves.

Finally, we stress that in our information-theoretic analysis of entanglement, we did not have to use concepts like spatial separation between subsystems of the composite system or the relative times of the observations on the subsystems. This is in agreement with Schrödinger's idea (c).

---

[2] Since the set of propositions entering Eq. (6.7) contains also propositions which are not mutually complementary for a non-product state, the sum in Eq. (6.7) does not give the value of information in bits contained in correlations for such states. It is then not surprising that, for example, in the case of a three-qubit GHZ state, the sum results in 4 ($I_{xyy} = I_{yxy} = I_{yyx} = I_{xxx} = 1$, the rest are zero). Here, however, only three of the four propositions are independent, as only 3 bits of information are available. This again is also a direct consequence of the formalism of quantum mechanics as the joint eigenstate of $\sigma_x^1 \sigma_y^2 \sigma_y^3$, $\sigma_y^1 \sigma_x^2 \sigma_y^3$ and $\sigma_y^1 \sigma_y^2 \sigma_x^3$ is also an eigenstate of $\sigma_x^1 \sigma_x^2 \sigma_x^3 = -(\sigma_x^1 \sigma_y^2 \sigma_y^3)(\sigma_y^1 \sigma_x^2 \sigma_y^3)(\sigma_y^1 \sigma_y^2 \sigma_x^3)$. Nevertheless, when the sum in Eq. (6.7) is larger than unity, this indicates that the total information contained in correlations is larger than one bit.

Up to this point, we did not specify any particular measure of information. Thus the question arises which particular measure of information is adequate to define the information gain in an individual quantum experiment.

Quantifying Information

Consider an experiment with two outcomes "yes" and "no" and with the probabilities p_1 and $p_2 = 1 - p_1$ respectively for the two outcomes. Within finite time, the experimenter can perform only a finite number of experimental trials. Because of inherent fluctuations associated with any probabilistic experiment with a finite number of trials, the number of occurrences of a specific outcome in future repetitions of the experiment is not precisely predictable. Rather, it obeys the binomial distribution (see e.g. Gnedenko 1976).

If one bets for example that the number of "yes" outcomes will be the one with highest probability, the probability of success still depends on p_1. With a probability of $p_1 = 0.5$, the probability of 5 "yes" outcomes in 10 trials is only 0.25, but with $p_1 = 0.9$ the probability of 9 "yes" outcomes in 10 trials is 0.39. It is a trait of the binomial distribution that we know the future number of occurrences of the outcomes very well if p_1 (or equivalently p_2) is close to 0 or 1, but we know much less about them when p_1 is around 0.5. Note that this follows from elementary probability theory without any input from physics.[3]

In Brukner and Zeilinger (1999) it was shown that this knowledge is represented by the measure

$$I = (p_1 - p_2)^2. \tag{6.8}$$

This attains its maximal value of unity when one of probabilities is one, and it attains its minimal value of 0 when both probabilities are equal.

Note that all our propositions about joint properties are binary propositions, i.e. they are associated with experiments with two possible outcomes, one of them being

---

[3] Here, a very subtle position was assumed by von Weizäcker (1975) who wrote: "It is most important to see that this [the fact that probability is not a prediction of the precise value of the relative frequency] is not a particular weakness of the objective empirical use of the concept of probability, but a feature of the objective empirical use of any quantitative concept. If you predict that some physical quantity, say a temperature, will have a certain value when measured, this prediction also means its expectation value within a statistical ensemble of measurements. The same statement applies to the empirical quantity called relative frequency. But here are two differences which are connected to each other. The first difference: In other empirical quantities, the dispersion of the distribution is in most cases an independent empirical property of the distribution and can be altered by more precise measurements of other devices; in probability the dispersion is derived from the theory itself and depends on the absolute number of cases. The second difference: In other empirical quantities the discussion of their statistical distributions is done by another theory than the one to which they individually belong, namely by the general theory of probability; in probability this discussion evidently belongs to the theory of this quantity, namely of probability itself. The second difference explains the first one."

"the product of spin of particle 1 along x_1, spin of particle 2 along x_2, ...and spin of particle N along x_N is +1", and the other "the product of spin of particle 1 along x_1, spin of particle 2 along x_2, ...and spin of particle N along x_N is -1", so that Eq. (6.8) can be applied. If we now denote the probabilities for the two outcomes by $p^+_{x_1,...,x_N}$ and $p^-_{x_1,...,x_N}$ respectively, then the information contained in proposition (*) is given by

$$I_{x_1,...,x_N} = (p^+_{x_1,...,x_N} - p^-_{x_1,...,x_N})^2. \tag{6.9}$$

Now we will express Eq. (6.9) in terms of the density matrix ρ of N qubits. First note that an arbitrary mixed state of N qubits can be written as

$$\rho = \frac{1}{2^N} \sum_{x_1,...,x_N=0}^{3} T_{x_1,...,x_N} \, \sigma^1_{x_1} \otimes, \ldots, \otimes \sigma^N_{x_N} \tag{6.10}$$

where σ^j_0 is the identity operator in the Hilbert space of particle j, and $\sigma^j_{x_j}$ is a Pauli operator for $x_j = 1, 2, 3$. Here, the elements of the correlation tensor T are given as mean values of the product of the N spins,

$$T_{x_1,...,x_N} = \text{Tr}[\rho(\sigma^1_{x_1} \otimes, \ldots, \otimes \sigma^N_{x_N})] = p^+_{x_1,...,x_N} - p^-_{x_1,...,x_N} \tag{6.11}$$

with $T_{0,...,0} = 1$. Then obviously our measure of information (Eq. 6.9) is equal to the square of the corresponding element of the correlation tensor

$$I_{x_1,...,x_N} = T^2_{x_1,...,x_N}. \tag{6.12}$$

We have thus obtained a quantitative expression for the individual measures of information contained in the sum of Eq. (6.7) and we want to emphasize that our analysis of entanglement would not be possible without the use of the measure of information Eq. (6.8).

So far in the present paper, we followed Schrödinger's concepts of entangled states in our information-theoretic analysis. Now we will follow Bell's ideas in a second, independent approach to characterize entanglement.

All Tight Bell-CHSH-Type Inequalities for Correlations of N Qubits

Here we present a single general Bell inequality that summarizes all possible constraints on statistical correlations of an N-qubit system. These constraints are derived under the assumptions of local realism. We consider such correlation measurements where for each individual particle one of two arbitrary dichotomic observables can be chosen. From this inequality we obtain as specific corollaries the Clauser-Horne-

Shimony-Holt (CHSH) inequality (Clauser et al. 1969) for two-qubit systems and the related inequalities for N qubits (Mermin 1990; Ardehali 1992; Belinskii and Klyshko 1993). The inequality is equivalent to the full (i.e. complete) set of all possible *tight* Bell-CHSH-type inequalities for N qubits.

In a local realistic picture, one assumes that the result of every measurement of an observable is predetermined.[4] Thus, in such a picture, one implicitly requires an unlimited amount of information to be carried by an individual particle, which conflicts with Schrödinger's idea (a).

Take an individual observer, denoted by index j, and allow him or her to be able to choose between two dichotomic observables (determined by some parameters denoted here $\mathbf{n}_1$ and $\mathbf{n}_2$). This implies the existence of two numbers $A_j(\mathbf{n}_1)$ and $A_j(\mathbf{n}_2)$ each taking values +1 or -1 which describe the predetermined result of a measurement by the observer of the observable defined by the local parameter $\mathbf{n}_1$ and $\mathbf{n}_2$, respectively. We choose such a notation for brevity; of course each observer can choose two arbitrary directions independently.

In a specific run of the experiment, the correlations between all N observations can be represented by the product $\prod_{j=1}^{N} A_j(\mathbf{n}_{k_j})$, with $k_j = 1, 2$. The correlation function then is the average over many runs of the experiment

$$E(\mathbf{n}_{k_1}, \ldots, \mathbf{n}_{k_N}) = \left\langle \prod_{j=1}^{N} A_j(\mathbf{n}_{k_j}) \right\rangle_{avg}. \tag{6.13}$$

Note that for each observer j, one has either $|A_j(\mathbf{n}_1) + A_j(\mathbf{n}_2)| = 0$ and $|A_j(\mathbf{n}_1) - A_j(\mathbf{n}_2)| = 2$ or the other way around. Then for all sign sequences of $s_1, \ldots, s_N$, where $s_j \in \{-1, 1\}$ the modulus of the product $|\prod_{j=1}^{N}[A_j(\mathbf{n}_1) + s_j A_j(\mathbf{n}_2)]|$ vanishes, except just one for which the product is 2^N. Therefore one has

$$\sum_{\substack{s_1,\ldots,s_N \\ =-1,1}} \left| \prod_{j=1}^{N}[A_j(\mathbf{n}_1) + s_j A_j(\mathbf{n}_2)] \right| = 2^N. \tag{6.14}$$

It then follows directly that the correlation functions must satisfy the following general Bell inequality (Weinfurter and Żukowski 2001; Żukowski and Brukner 2002; for an independent derivation, see Werner and Wolf 2001)

$$\sum_{\substack{s_1,\ldots,s_N \\ =\pm 1}} \left| \sum_{\substack{k_1,\ldots,k_N \\ =0,1}} s_1^{k_1}, \ldots, s_N^{k_N} E(\mathbf{n}_{k_1}, \ldots, \mathbf{n}_{k_N}) \right| \leq 2^N. \tag{6.15}$$

Therefore within each modulus we have sums of all 2^N correlation functions, which are the result of the Bell-type experiment. However, each correlation function is

---

[4] We do not discuss local stochastic hidden variable theories explicitly, as any such model can be constructed from an underlying deterministic one. The bounds of Bell inequalities are reached by the deterministic theories.

multiplied by a specific sign $\pm$. Using a generalization of the fact that for a pair of real numbers a and b one always has $|a \pm b| \le |a| + |b|$, one can obtain the set of Bell-type inequalities which as a whole are equivalent to inequality (6.15)

$$|\sum_{\substack{s_1,\dots,s_N \\ =-1,1}} S(s_1,\dots,s_N) \sum_{\substack{k_1,\dots,k_N \\ =0,1}} s_1^{k_1},\dots,s_N^{k_N} E(\mathbf{n}_{k_1},\dots,\mathbf{n}_{k_N})| \le 2^N, \qquad (6.16)$$

where $S(s_1,\dots,s_N)$ is one of the 2^{2^N} sign functions of $s_1,\dots,s_N$, by which we mean that its possible values can only be $+1$ or -1. Actually, inequalities (6.16) represent the complete set of tight inequalities of the CHSH-type (for binary observables, two observables per observer, involving only N-system correlation functions) for N observers[5] of all possible 2^{2^N} inequalities for the correlations, one for each possible choice of the sign function S. Many of these inequalities are trivial (for example when the choice is $S(s_1,\dots,s_N) = 1$ for all arguments, we get that the modulus of the correlation function does not exceed 1). An example: take $S(s_1,\dots,s_N)$ given by $\sqrt{2}\cos(-\frac{\pi}{4} + \frac{\pi}{2}\sum_{j=1}^{N} s_j)$. This leads to the series of inequalities derived by Belinskii and Klyshko (1993). Specifically, for $N = 2$, the CHSH inequality

$$|E(1, 1) + E(1, 2) + E(2, 1) - E(2, 2)| \le 2 \qquad (6.17)$$

follows. For $N = 3$, one obtains

$$|E(1, 2, 2) + E(2, 1, 2) + E(2, 2, 1) - E(1, 1, 1)| \le 2, \qquad (6.18)$$

where here we use numbers 1 and 2 to denote directions $\mathbf{n}_1$ and $\mathbf{n}_2$, respectively. Inequality (Eq. 6.18) leads to the Greenberger-Horne-Zeilinger contradiction (Greenberger et al. 1989, 1990) for an appropriate choice of local settings. In these cases, the left-hand side of inequality (6.18) reaches the value 4, which is the maximum possible value for any, not only quantum, correlation function.

Thus far we have shown that when a local realistic model applies, the general Bell inequality (6.15) follows. The reverse is also true and we give the proof below: whenever inequality (6.15) holds, one can construct a local realistic model for the correlation function values for the studied problem (N systems, binary observables, two measurement settings per observer). This establishes the general Bell inequality presented above as a necessary and sufficient condition for local realistic description of multi-particle correlations, where two dichotomic observables are measured on each individual particle. Additionally this proves that the set of tight inequalities (Eq. 6.16) is complete, as no other inequality is needed to bound the full set of possible local hidden variable models of correlation functions for the studied problem.

The proof of the sufficiency of condition (6.15) will be done in a constructive way. One simply ascribes to the set of predetermined local results, which satisfy the

---

[5] For an extensive classification of the inequalities see (Werner and Wolf 2001). For three qubits, a complete set of inequalities has been found numerically in Pitowsky and Svozil (2001). See also Pitowsky (1989) and Peres (1999).

following conditions $A_j(\mathbf{n}_1) = s_j A_j(\mathbf{n}_2)$, the hidden probability $p(s_1, \ldots, s_N) = \frac{1}{2^N} |\sum_{k_1, \ldots, k_N} s_1^{k_1}, \ldots, s_N^{k_N} E(\mathbf{n}_{k_1}, \ldots, \mathbf{n}_{k_N})|$, and one demands that the product $\prod_{j=1}^{N} A_j(\mathbf{n}_2)$ has the same sign as that of the expression inside of the modulus defining $p(s_1, \ldots, s_N)$. In this way a definite set of local realistic values is ascribed a unique global hidden probability. Obviously, such defined probabilities are positive. However due to inequality (6.15), they may add up to less than 1. In this case, the "missing" probability is ascribed to an arbitrary model of local realistic noise (e.g., for which all possible products of local results enter with equal weights). The overall contribution of such a noise term to the correlation function is zero. In this way, we obtain a local realistic model of a certain correlation function. However, one should check that this construction indeed reproduces the model for the correlation function for the set of settings that enter inequality (6.15), that is for $E(\mathbf{n}_{k_1}, \ldots, \mathbf{n}_{k_N})$. For simplicity take $N = 2$. Notice that the expansion coefficients of the four-dimensional vector $(E(\mathbf{n}_1, \mathbf{n}_1), E(\mathbf{n}_1, \mathbf{n}_2), E(\mathbf{n}_2, \mathbf{n}_1), E(\mathbf{n}_2, \mathbf{n}_2))$ in terms of orthogonal basis vectors $(s_1 s_2, s_1^2 s_2, s_1 s_2^2, s_1^2 s_2^2)$ (recall that $s_1, s_2 \in \{-1, 1\}$) are equal to the expressions within the moduli entering inequality (6.15). Next, notice that by the construction shown above, the local realistic model for $N = 2$ gives

$$(E_{LV}(\mathbf{n}_1, \mathbf{n}_1), E_{LV}(\mathbf{n}_1, \mathbf{n}_2), E_{LV}(\mathbf{n}_2, \mathbf{n}_1), E_{LV}(\mathbf{n}_2, \mathbf{n}_2))$$

$$= \frac{1}{4} \sum_{s_1, s_2} \sum_{k_1, k_2} s_1^{k_1} s_2^{k_2} E(\mathbf{n}_{k_1}, \mathbf{n}_{k_2})(s_1 s_2, s_1^2 s_2, s_1 s_2^2, s_1^2 s_2^2).$$

Thus, since the vectors built out of the correlation function values and its local realistic counterpart have the same expansion coefficients, they are equal and the sufficiency of (6.15) as a condition for local realism is proven. The generalization to an arbitrary N is obvious.

Above, we derived the full set of Bell inequalities for multi-qubit correlations. This strictly defines the boundary of the validity of local realism. We will now discuss states which violate such inequalities with the specific aim of investigating the way information can be distributed between the subsystems of such states.

N Qubits that Violate Local Realism

Let us consider the general N-qubit state as in Eq. (6.10). Then, the N-qubit quantum correlation function for a Bell-GHZ type experiment is

$$E_{QM}(\mathbf{n}_{k_1}, \ldots, \mathbf{n}_{k_N}) = \mathrm{Tr}[\rho(\boldsymbol{\sigma} \cdot \mathbf{n}_{k_1} \otimes, \ldots, \otimes \boldsymbol{\sigma} \cdot \mathbf{n}_{k_N})] \qquad (6.19)$$

$$= \sum_{x_1,\ldots,x_n=1}^{3} T_{x_1,\ldots,x_N}(\mathbf{n}_{k_1})_{x_1}, \ldots, (\mathbf{n}_{k_N})_{x_N} \qquad (6.20)$$

where $(\mathbf{n}_{k_j})_{x_j}$ $(x_j = 1, 2, 3)$ are the three Cartesian components of the vector $\mathbf{n}_{k_j}$. Equation (6.19) means that the N-particle correlation function is fully defined by a tensor $\hat{T}$ (the indices of which can take values 1, 2, 3, and which belongs to R^{3N}). For convenience we shall write down the last equation in a more compact way as $E_{QM}(\mathbf{n}_{k_1}, \ldots, \mathbf{n}_{k_N}) = \langle \hat{T}, \mathbf{n}_{k_1} \otimes, \ldots, \otimes \mathbf{n}_{k_N} \rangle$, where $\langle \ldots, \ldots \rangle$ denotes the scalar product in R^{3N}.

The necessary and sufficient condition (6.15) for a local realistic description of N-particle correlations implies that the quantum correlations for N qubits can always have a local and realistic model for the Bell-type experiment if and only if

$$\sum_{\substack{s_1,\ldots,s_N \\ =-1,1}} |\langle \hat{T}, \sum_{k_1=1}^{2} s_1^{k_1} \mathbf{n}_{k_1} \otimes, \ldots, \otimes \sum_{k_N=1}^{2} s_N^{k_N} \mathbf{n}_{k_N} \rangle| \leq 2^N \qquad (6.21)$$

for any possible choice $\mathbf{n}_{k_1}, \ldots, \mathbf{n}_{k_N}$ of each observer's two local settings $\mathbf{n}_1$ and $\mathbf{n}_2$.

This condition can be simplified further, by noticing that for each observer there always exist two mutually orthogonal unit vectors $\mathbf{a}_1$ and $\mathbf{a}_2$, independently defined for each observer, and the angle α_j such that $\sum_{k_j=1}^{2} \mathbf{n}_{k_j} = 2\mathbf{a}_1 \cos(\alpha_j + \frac{\pi}{2})$ and $\sum_{k_j=1}^{2} (-1)^{k_j} \mathbf{n}_{k_j} = 2\mathbf{a}_2 \cos(\alpha_j + \pi)$. Let us put $c_{x_j} = \cos(\alpha_j + x_j \frac{\pi}{2})$. Then, one can write the inequality (6.21) as

$$\sum_{\substack{x_1,\ldots,x_N \\ =1,2}} |c_{x_1}, \ldots, c_{x_N} \langle \hat{T}, \mathbf{a}_{x_1} \otimes, \ldots, \otimes \mathbf{a}_{x_N} \rangle| \leq 1. \qquad (6.22)$$

One can rewrite this inequality as

$$\sum_{\substack{x_1,\ldots,x_N \\ =1,2}} |c_{x_1}, \ldots, c_{x_N} \langle \hat{T}, \mathbf{a}_{x_1} \otimes, \ldots, \otimes \mathbf{a}_{x_N} \rangle| \leq 1 \qquad (6.23)$$

where $T_{x_1,\ldots,x_N}$ is now a component of the tensor $\hat{T}$ in a new set of local coordinate systems, which among their basis vectors have $\mathbf{a}_1$ and $\mathbf{a}_2$ which serve as the unit vectors which define the directions x and y.

The necessary and sufficient condition for impossibility of any local realistic description of N-qubit correlations is that the maximum of the left-hand side of inequality (6.23) is larger than one. Once the values of the elements of the correlation tensor are given for the specific density matrix one can check via maximization procedure whether the local realistic description is possible.

On the other hand, it is easy to notice that with the aid of the Cauchy-Schwarz inequality, one has an inequality

$$\sum_{\substack{x_1,...,x_N \\ =1,2}} |c_{x_1}, \ldots, c_{x_N} T_{x_1,...,x_N}| \leq \sqrt{\sum_{\substack{x_1,...,x_2 \\ =1,2}} T_{x_1,...,x_2}^2}. \qquad (6.24)$$

Combining inequality (6.23) with (6.24) we obtain that a sufficient condition for the possibility of a local realistic description of the quantum N-qubit correlations in any Bell-type experiment is that

$$\sum_{\substack{x_1,...,x_N \\ =1,2}} T_{x_1,...,x_N}^2 \leq 1 \qquad (6.25)$$

for all possible choices of local coordinate systems for N qubits, as then the full set of Bell inequalities (6.16) is satisfied. Equivalently, if at least one of the Bell inequalities from the set is violated, then condition (6.25) is violated, for at least one choice of local coordinate systems.

Our measure of information (6.12) is exactly equal to the square of the corresponding element of the quantum correlation matrix. This establishes the equivalence between the condition for local realism (6.25) on correlation tensor elements and our information-theoretical criterion (6.7) for information contained in correlations by states which do not reveal quantum entanglement. If the state violates at least one of the Bell inequalities from the full set (6.16) then this state is characterized by the information-theoretical criterion (6.5).

By performing rotations in the x-y planes of the N observers, one can vary the values of the elements of the correlation tensor, but these variations do not change the left-hand side of inequality (6.25). In information-theoretic language we say that the total information content in x-y plane correlations is invariant under these variations. The invariance property implies that one can find local coordinate systems for which some of the correlation tensor elements vanish, thus arriving at criterion (6.25) which involves a smaller number of them [for the case of three qubits see Scarani and Gisin (2001)]. For example, in the two-qubit case, the rotations in the x-y planes of the two observers are obtained with the use of two parameters, each describing the rotation angle for the given local observer, and therefore one can always find local coordinate systems such that two of correlation tensor elements vanish ($T_{xy} = T_{yx} = 0$) [see Horodecki and Horodecki (1996)]. Then it can easily be seen that varying the two angles α_1 and α_2, the expression on the left-hand side of inequality (6.24) can be saturated by the one on the right-hand side. Finally, this establishes condition (6.25) for two qubits as the necessary and sufficient condition for the correlations measured on an arbitrary two-qubit mixed state to be understood within the local realistic picture. In that case, our necessary and sufficient condition $I_{xx} + I_{yy} > 1$ (or equivalently $T_{xx}^2 + T_{yy}^2 > 1$) for violation of the most general Bell's inequality is equivalent with the necessarily and sufficient condition for violation of the CHSH inequality which was obtained by the Horodecki et al. (1995). Thus, our result also confirms that non-violation of the CHSH inequality is a necessary and sufficient condition for the local realistic description of two-qubit correlations.

We will now analyze from our information-theoretic perspective the case of an N-qubit Bell-type experiment. We are specifically interested in the limit up to which the experiment still has a local realistic interpretation. Consider a state which is a mixture of the maximally entangled state and the noise induced by experimental imperfections. Such a state is known as the Werner state and has the form

$$\rho_W = V|\psi_{GHZ}\rangle\langle\psi_{GHZ}| + (1-V)\rho_{noise} \tag{6.26}$$

where $|\psi_{GHZ}\rangle = \frac{1}{\sqrt{2}}(|+z\rangle_1 \ldots |+z\rangle_N + |-z\rangle_1 \ldots |-z\rangle_N)$ is the maximal entangled (GHZ) state and $\rho_{noise} = \frac{1}{2^N}I$ is the completely mixed state. Here, e.g., $|+z\rangle_j$ denotes the spin up of particle j along z. We would like to emphasize that the weight V of the GHZ-state can be interpreted as the visibility observed in a multi-particle interference experiment (Belinskii and Klyshko 1993).

For the purposes of our argument we will now calculate the number of non-zero correlation tensor elements (which are related to our individual measures of information contained in the correlations) for the Werner state. Note first that for any measurement direction $\mathbf{n}$ belonging to the x-y plane the spin component $\mathbf{n} \cdot \boldsymbol{\sigma}$ has its eigenvectors in the form $|\pm \mathbf{n}\rangle = \frac{1}{\sqrt{2}}(|+z\rangle \pm e^{i\phi}|-z\rangle)$, where ϕ is the azimuthal angle of the vector $\mathbf{n}$. Using Eq. (6.19) one can easily show that the correlation function for arbitrary chosen measurement directions within the local x-y planes is

$$E_W(\phi_1, \ldots, \phi_N) = V \cos(\sum_{k=1}^{N} \phi_k). \tag{6.27}$$

This implies that the correlation tensor elements $T_{x_1,\ldots,x_N}$ with $x_1, \ldots, x_N$ each being either x or y are given by $T_{x_1,\ldots,x_N} = V \cos(m_y \frac{\pi}{2})$, where m_y is the number of y's in $\{x_1, \ldots, x_N\}$. Therefore, for each N, one always has $T_{xx\ldots x} = V$. The other components are zero, except for those that contain an even number of y's, which are either V or $-V$. This results in the total number $2 + \sum_{k=1}^{N-1} \binom{N}{N-2k} = 2^{N-1}$ of non-zero components for the even N and $1 + \sum_{k=1}^{(N-1)/2} \binom{N}{N-2k} = 2^{N-1}$ for the odd N.

We would like to stress again the equivalence $I_{x_1,\ldots,x_N} = T^2_{x_1,\ldots,x_N}$ between the measure of information contained in the correlation between measurements performed along directions $x_1, \ldots, x_N$ and the square of the corresponding correlation tensor element. In the case of N qubits in the Werner state we therefore have 2^{N-1} individual measures of information with the value V^2 and we have the remaining ones equal to zero. Inserting these values into the information criterion (6.7) [or equivalently (6.25)] one obtains

$$V \le \left(\frac{1}{\sqrt{2}}\right)^{N-1} \tag{6.28}$$

for the maximal visibility which still allows the correlations between N qubits in the Werner state to be understood within a local realistic picture. Note that in such a case the right-hand side of the inequality (6.24) is one. The value given on the right-hand side of (6.28) is also the minimal visibility necessarily to violate the inequalities for N qubits obtained by Belinskii and Klyshko (1993). Since these inequalities are included in our set of all possible inequalities (6.16) we can conclude that our information criterion (6.7) is the necessary and sufficient condition for correlations between N qubits in the Werner state to violate the local realistic description.

It is interesting to note that, in Żukowski (1993), an inequality was obtained for correlations to be understood within local realism where not only two but all possible measurement settings are chosen by N observers. There, for N>3 an even lower threshold for the visibility was obtained to violate this inequality. For such an experimental situation, our criterion (6.7) is the sufficient condition for violation of the inequality.

Another interesting observation is that for N maximally entangled qubits, N bits of information rest in the correlations, as opposed to always not more than one bit for the classically composed ones. In the case of GHZ (for $V = 1$ in the consideration given above) the information criterion (6.7) results in $2^{N-1} \leq 1$ which clearly shows that with growing N the discrepancy between quantum and classical correlations grows exponentially. This is in concurrence with the fact that the GHZ theorem is stronger than Bell's and its strength, as measured by the magnitude of violation of (6.16) for maximally entangled states, exponentially increases with the number of qubits (Mermin 1990; Ardehali 1992; Belinskii and Klyshko 1993).

Conclusions

We now would like to review what we have done in the present paper and place it in a broader perspective. The paper contains two independent main approaches to the question of quantum entanglement, and we finally show their essential equivalence.

In the first approach we start from the conceptual position that quantum mechanics is about information. We express the information contained in composite systems such that it can be divided into the information carried by the individuals versus the information contained in the correlations between observations made on the individuals. We further assume that the information contained in any system, be it individual or composite, is finite.

Considering first the classically composed systems, we note that any correlations we might observe between the subsystems of such a composite system can simply be understood on the basis of correlations between the properties the individual subsystems have on their own. This means that if we know all properties of the individual subsystems, we can definitely conclude how much information is contained in their correlations. For quantum entangled systems, this is not true anymore. Such composite quantum systems can carry more information in joint properties than what may be concluded from knowledge of the individuals. These considerations lead to a natural information-based understanding of quantum entanglement. Within this view we see Mermin's "correlations without correlata" (Mermin 1998) as reflecting that when correlations are defined there is no information left to define "correlata" as well. In this case, "correlations have physical meaning; that which they correlate does not" (Mermin 1998).

In an independent approach, we obtain the most general set of Bell inequalities for N systems and dichotomic observables. That way, we arrive at a necessary and sufficient condition for N qubits whose correlations cannot be understood within local realism. Local realism is based on the assumption that results of the observations on the individual systems are predetermined and independent of whatever measurements might be performed distantly. One may notice that this assumption implicitly says that correlations between subsystems do not go beyond what might be concluded from the properties of individual subsystems.

We finally show that the two approaches, the information theoretical one and the one via Bell's inequalities, are equivalent in their essence. This is done via the fact that the Bell inequality criteria can be translated into a statement about correlations (probabilities), which again can be understood as an information theoretical expression. This requires the use of a new measure of information introduced earlier (Brukner and Zeilinger 1999). This measure of information is distinct from Shannon's measure (Shannon 1948). The main conceptual difference is that Shannon's measure tacitly assumes that the properties of the systems carrying the information are already well defined prior to, and independent of, observation (Brukner and Zeilinger 2001). In quantum mechanics this clearly is not the case. There, the criterion for choosing the new measure of information was that it is invariant on the experimentalist's free choice of a complete set of mutually complementary observables.

In conclusion, we would like to draw the reader's attention to the fact of the equivalence of the two approaches in the present paper, the information theoretic one and the one via Bell's inequalities. It is evident that the first one is much more simple both conceptually and formally. It is suggestive that this new information theoretic formulation of quantum phenomena opens up the avenue of new approaches to well known problems in quantum information physics and in the foundations of quantum mechanics.

Further Developments

The information-theoretic approach to quantum entanglement, which was developed 19 years ago in this paper (ibid.), has contributed to an overall growth of new results on information theoretic foundations of quantum physics and inspired an increasing number of scientists joining the general research direction. As a consequence, a diversity of reconstructive approaches to quantum theory have emerged over the last decade with the aim of explaining the physical content of the theory by reconstructing it from a set of simple, plausible information-theoretical principles (Hardy 2016; Spekkens 2007; Dakić and Brukner 2009, 2016; Chiribella et al. 2011; Caticha 2011; Masanes and Müller 2011; Höhn and Wever 2017; Höhn 2017; Pawlowski et al. 2009). In parallel, these approaches have made it possible to understand the key underlying physical ideas that provide deeper insight into the nature of quantum reality. In turn, their explanatory power enabled designing novel tests of the foundations of quantum theory within hitherto unachieved parameter regimes and with unprecedented levels of accuracy.

Acknowledgements A.Z. acknowledges funding support from the Austrian Academy of Sciences and from the University of Vienna via the project QUESS. M.Ż. acknowledges the ICTQT IRAP (MAB) project of FNP, co-financed by structural funds of the EU.

References

Ardehali M (1992) Bell inequalities with a magnitude of violation that grows exponentially with the number of particles. Phys Rev A 46:5375–5378

Aspect A, Grangier P, Roger G (1981) Experimental tests of realistic local theories via Bell's theorem. Phys Rev Lett 47:460–463

Belinskii AV, Klyshko DN (1993) Interference of light and Bell's theorem. Phys Usp 36:653–693

Bell JS (1964) On the Einstein-Podolsky-Rosen paradox. Physics 1, 195–200. Reprinted in Bell JS (1987) Speakable and unspeakable in quantum mechanics. Cambridge University Press

Bohr N (1958) Atomic physics and human knowledge. Wiley, New York

Brukner Č, Zeilinger A (1999) Operationally invariant information in quantum measurements. Phys Rev Lett 83:3354–3357

Brukner Č, Zeilinger A (2001) Conceptual inadequacy of the Shannon information in quantum measurements. Phys Rev A 63:022110–022113

Caticha A (2011) Entropic dynamics, time and quantum theory. J Phys A 4:22

Chiribella G, D'Ariano T, Simon GM, Perinotti P (2011) Informational derivation of quantum theory. Phys Rev A **84**, 012311

Clauser J, Horne M, Shimony A, Holt R (1969) Proposed experiment to test local hidden-variable theories. Phys Rev Lett 23:880–884

Dakić B, Brukner Č (2009) Deep beauty: understanding the quantum world through mathematical innovation. In: Halvorson E (ed) Quantum theory and beyond: is entanglement special?. Cambridge University Press, Cambridge, pp 365–392 arXiv:0911.0695 [quant-ph]

Dakić B, Brukner Č (2016) Quantum theory: informational foundations and foils. In: Chiribella G, Spekkens R (eds) The classical limit of a physical theory and the dimensionality of space. Springer, Heidelberg, pp 249–282 arXiv:1307.3984 [quant-ph]

Einstein A, Podolsky B, Rosen N (1935) Can quantum-mechanical description of physical reality be considered complete? Phys Rev 47:777–780

Freedman SJ, Clauser JS (1972) Experimental test of local hidden-variable theories. Phys Rev Lett 28:938–941

Gisin N (1991) Bell's inequality holds for all non-product states. Phys Lett A 154:201–202

Gisin N, Peres A (1992) Maximal violation of Bell's inequality for arbitrary large spin. Phys Lett A 162:15–17

Gnedenko BV (1976) The theory of probability. Mir Publishers, Moscow

Greenberger DM, Horne M, Zeilinger A (1989) Bell's Theorem, quantum theory, and conceptions of the Universe. In: Kafatos M (ed) Going beyond Bell's Theorem. Kluwer Academic, Dordrecht, pp 69–72

Greenberger DM, Horne M, Zeilinger A (1993) Multiparticle interferometry and the superposition principle. Phys Today 46:22–29

Greenberger DM, Horne M, Shimony A, Zeilinger A (1990) Bell's theorem without inequalities. Am J Phys 58:1131–1143

Hardy L (2016) Quantum theory: informational foundations and foils. In: Chiribella G, Spekkens R (eds) Reconstructing quantum theory. Springer, Heidelberg, pp 223–248

Höhn PA (2017) Toolbox for reconstructing quantum theory from rules on information acquisition. Quantum 1:38

Höhn PA, Wever CSP (2017) Quantum theory from questions. Phys Rev A 95:012102

Horne MA, Zeilinger A (1985) A Bell-Type EPR experiment using linear momenta. In: Symposium on the foundations of modern physics. Scientific Publ, Singapore, pp 435–443

Horodecki R, Horodecki M (1996) Information-theoretic aspects of inseparability of mixed states. Phys Rev A 54:1838–1843

Horodecki R, Horodecki P, Horodecki M (1995) Violating Bell inequality by mixed spin-1/2 states: necessary and sufficient condition. Phys Lett A 200:340–344

Kaszlikowski D, Gnacinski P, Żukowski M, Miklaszewski W, Zeilinger A (2000) Violations of local realism by two entangled N-dimensional systems are stronger than for two qubits Phys Rev Lett **85**:4418–4421

Masanes L, Müller M (2011) A derivation of quantum theory from physical requirements. New J Phys 13:063001

Mermin ND (1990) Extreme quantum entanglement in a superposition of macroscopically distinct states. Phys Rev Lett 65:1838–1841

Mermin ND (1998) What is quantum mechanics trying to tell us? Am J Phys 66:753–767

Pan JW, Bouwmeester D, Weinfurter H, Zeilinger A (2000) Experimental test of quantum nonlocality in three photon Greenberger-Horne-Zeilinger entanglement. Nature 403:515–518

Pawlowski M, Paterek T, Kaszlikowski D, Scarani V, Winter A, Żukowski M (2009) Information causality as a physical principle. Nature 461:1101–1104

Peres A (1999) All the Bell inequalities. Found Phys 29:589–614

Pitowsky I (1989) Quantum probability—quantum logic. Springer, Berlin

Pitowsky I, Svozil K (2001) Optimal tests of quantum nonlocality. Phys Rev A 64:014102

Popescu S (1995) Bell's inequalities and density matrices: revealing 'hidden' nonlocality. Phys Rev Lett 74:2619–2622

Rauch H, Zeilinger A, Badurek G, Wilfing A, Bauspiess W, Bonse U (1975) Verification of Coherent Spinor Rotation of Fermions. Phys Lett A 54:425–427

Scarani V, Gisin N (2001) Spectral decomposition of Bell's operators for qubits. J Phys A 34:6043

Schrödinger E (1935a) Die gegenwärtige Situation in der Quantenmechanik. Naturwissenschaften **23**:807–812823828. Translation published in Proc Am Phil Soc **124**:323–338 and in Wheeler JA, Zurek WH (eds) Quantum theory and measurement. Princeton University Press, Princeton, pp 152–167. A copy can be found at ⟨www.emr.hibu.no/lars/eng/cat⟩

Schrödinger E (1935b) Discussion of probability relations between separated systems. Proc Camb Philos Soc 31:555

Shannon CE (1948) A mathematical theory of communication. Bell Syst Tech J **27**:379. A copy can be found at ⟨http://cm.bell-labs.com/cm/ms/what/shannonday/paper.html⟩

Spekkens RW (2007) Evidence for the epistemic view of quantum states: a toy theory. Phys Rev A 75:032110

von Weizsäcker CF (1975) Quantum theory and the structures of time and space II. In: Castell MDL, von Weizsäcker CF (eds)The philosophy of alternatives. Hanser, Munich, pp 213–230. Papers presented at a conference held in Feldafing, July 1974

Weihs G, Jennewein T, Simon C, Weinfurter H, Zeilinger A (1998) Violation of Bell's inequality under strict Einstein locality conditions. Phys Rev Lett 81:5039–5043

Weinfurter H, Żukowski M (2001) Four-photon entanglement from down-conversion. Phys Rev A 64:010102

Werner RF, Wolf MM (2001) All multipartite Bell correlation inequalities for two dichotomic observables per site. Phys Rev A **64**:032112. Preprint quant-ph/0102024 (2001)

Zeilinger A (1997) Quantum teleportation and the non-locality of information. Phil Trans Roy Soc Lond 1733:2401–2404

Zeilinger A (1999) A foundational principle for quantum mechanics. Found Phys 29:631–643

Żukowski M (1993) Bell Theorem involving all settings of measuring apparata. Phys Lett A 177:290

Żukowski M, Brukner V (2002) Bell's theorem for general N-qubit states. Phys Rev Lett **88**:210401. Preprint quant-ph/0102039 (2001)

Chapter 7
Does a Single Lone Particle Have a Wavefunction ψ? YES. An Experimental Test of Einstein's "Unfinished Revolution"

Herbert J. Bernstein

Introduction and Reminiscences

Michael A Horne was a wonderful colleague to all who worked with him. With me that was over more than three decades. For about twenty-one of those years I was a collaborator, serving as his Principal Investigator on a NSF-funded effort researching Quantum Interferometry. Granted to Hampshire College, this series of awards was the successor to Cliff Shull's MIT neutron diffraction and interference experimental program. It supported me, with Mike, Anton Zeilinger and Danny Greenberger as co-PIs and Cliff[1] as senior adviser right up to his death in 2001. As a group, the four of us asked ourselves "WHY the quantum?" Not how does it work, not where does ordinary reality come from, nor "How can the quantum possibly accord with reductionism," especially since the fundamental, most reduced, layers of objects— the level at which QM operates, things are so unpicturable, so inseparable, having no local-reality; particles are so unreal, seeming not to possess any definite properties until they are sought.[2] No; we asked "Why the quantum?" and for many years we

---

[1] Winner of the 1994 Nobel Prize in Physics, Cliff shared the honor with Bertram Brockhouse; Mike was officially subcontracted on the NSF grants through Stonehill College so they could participate in the concessionary overhead rate, too.

[2] Even John A Wheeler's intriguing search for the "*It* from the Bit" was much less a form of our question than a brilliant foreshadowing of the whole field of Quantum Information. A field we each were ideally positioned to enter, and did to varying degrees from several angles.

Electronic Supplementary Material The online version of this chapter (https://doi.org/10.1007/978-3-030-77367-0_7) contains supplementary material, which is available to authorized users.

H. J. Bernstein (✉)
Physics Professor & President of The IS&IS Institute for Science, School of NS, Hampshire College, Amherst, MA, USA
e-mail: hBernstein@hampshire.edu

© Springer Nature Switzerland AG 2021
G. Jaeger et al. (eds.), *Quantum Arrangements*, Fundamental Theories of Physics 203,
https://doi.org/10.1007/978-3-030-77367-0_7

met in coffee houses, at each-others' homes and in conference venues everywhere, and from New York to Vienna—with Amherst and Boston in between.

This special collaboration produced many breakthrough "firsts" in both experiment and theory.[3] While it started with Neutron Interferometry, having both authors of the two fundamental applications[4] that launched the field, it soon became a major research venue for all kinds of interferometry, and many aspects of quantum phenomena not thought by others to be interference at all. We even answered a few fundamental questions from the era of great debates on The Quantum and its interpretation. Can the 1935 definition of an *"element of reality"* in the famous EPR paper be made consistent? NO.[5] Since every complete "von Neumann" measurement corresponds to a Hermitian matrix: Can every Hermitian operator be made measurable? YES.[6]

Three Historical Points

Michael A Horne and I first met in April or May 1978 with Cliff at Shull's lab in the MIT reactor building. The effect Mike was working on for a conference presentation was the Fizeau effect, investigating phase shifts induced on neutron waves by travel in a moving medium.[7] I believe Anton Zeilinger was there too of course. It was almost immediately clear we all had a great interest in that salient "Why the quantum?" In honor of that meeting and of the decades-long collaboration and friendship which followed, about which more later, I hereby propose an experiment to test the "Statistical Interpretation" of QM, the last gasp objection Einstein held to in his long attempt to see behind the quantum veil.

---

[3] The collaboration produced hundreds of articles; at least a dozen sported co-authorship by me and Mike. Perhaps a score show Anton Zeilinger and me among their authors, fewer with Danny Greenberger's and my names attached. And there are even a couple where I appeared as co-author with Cliff himself.

[4] Greenberger's gravitational phase shift when the arms of a two-path massive particle interferometer are not at the same height in Earth's field was the first experiment demonstrating the quantum mechanical effect of a Gravity interaction: a fact underlined by the appearance of both "g" and $\hbar$ in the formula for observed beam intensities. My suggestion of a direct demonstration that for Fermions (all HALF-Integer Spin particles), an odd number of complete rotations about any given axis introduces an extra minus sign. A full 360° rotation shifts the phase by 180° & turns all points of constructive interference into destructive and *vice-versa*. Many experts (and textbooks of the time) stated unequivocally that the Born rule for squaring amplitudes meant this factor of -1 could never be observed.

[5] D. M. Greenberger, M. A. Horne, and A. Zeilinger, in Bell's Theorem, Quantum Theory, and Conceptions of the Universe, edited by M. Kafatos (Kluwer Academic, Dordrecht, 1989), p. 69 analyzed a 4 particle singlet state formed from 2 spin-½ triplets. See also N. David Mermin, Am. J. Phys.58, 731 (1990) who recognized the 3-particle form of the "GHZ Theorem".

[6] Experimental Realization of Any Discrete Unitary Operator, M. Reck, A. Zeilinger, H. J. Bernstein and P. Bertani, Phys. Rev. Lett. *73*, 58 (1994). https://doi.org/10.1103/PhysRevLett.73.58.

[7] Neutron Interferometry. Proceedings of a [VI/1978] workshop, Institut Max von Laue–Paul Langevin, Grenoble. U. Bonse, H. Rauch, eds Oxford Press 1979, p. 350.

This seems to be Albert Einstein's last stand, his position against acceptance of the quantum theory when he wrote or spoke the autobiographical notes[8] with Paul A Schilpp. On page 671 Einstein claims the wavefunction ψ only describes an ensemble of identically prepared individual systems, on the basis of what "statistical QM" cannot or does not provide. He asserted more was necessary to have not just a correct computation of results as far as it goes, but a truly *complete* theory for individual particles. The present paper provides a simple experiment to that demonstrate this demand is wrong; one does not need such a huge ensemble of identical preparations to get the results of the individual particle as predicted by its own wavefunction ψ.

The last historical point is more recent: A third incident made me think the notion that QM requires vast ensembles still holds some appeal. It happened to me at a conference in Vienna where all four co-PIs were present. During a plenary talk on my SuperDense teleportation[9] (SdT) I emphasized that Charles the chooser needs to give Alice only a SINGLE particle with wavefunction ψ for her to have Bob reproduce it unerringly in the remote laboratory. Even the original Quantum Teleportation (QT) works as a one-shot experiment. An excellent famous German scientist (who won the Nobel prize a few years later) came up after the talk—upon hearing my emphasis—and invited himself to lunch at my table.[10] His general philosophy includes a goodly dose of operationalism. Clearly the notion of an experiment involving the wavefunction of a single particle has some intellectual attraction. Is it possible that the statistical interpretation or, more precisely, the operational need for many copies of the same preparation still serves as the mental anodyne implicitly covering QM mysteries with the need for a vast ensemble? If so, the experiment proposed here may help cure that subtle coverage or addiction.

---

[8] In Einstein's own words, "the ψ-function is to be understood as the description not of a single system but of an ensemble of systems." Using this idea the seeming-paradoxes generated by the alternative conception, that the ψ-function is a complete description of the single system, were supposed to disappear. (PA Schilpp, *ed.* Autobiographical Notes by Albert Einstein, 1949. The quote is in a section intriguingly titled "Replies to Criticisms").

[9] SuperDense Quantum Teleportation by Remote State Preparation https://doi.org/10.1007/s11128-006-0030-5, Springer Verlag, (XII 2006).

[10] I guess when he heard that Diana (the deployer or final disposer of ψ) still had to check that the whole scheme works, he was satisfied that his initial conception of quantum strangeness—while a bit shaken & maybe even extended and supplemented with a new example—was still adequate. And he perhaps became less impressed, less eager to investigate by launching an SdT experiment himself. For Diana is key: in the run-Up to using the system for practical teleportations she would have to make Charles send her many copies from the category of wavefunction she wanted to measure. They are necessary to orient the apparatus and the transformations that Bob must use.

The Experiment

The experiment is inspired by one we use to teach "QM for the Myriad," a signature Hampshire course[11] which puts modern physics first and reverses the historical sequence introductory college physics usually uses. In the current instance I was inspired by an early photon-counting demonstration recorded on film nearly fifty years ago by Stefan Berko: "Polarization of Single Photons" [attached QuickTime movie, and/or http://tsgphysics.mit.edu/front/?page=demo.php&letnum=Z%2016& show=0]. It shows how a vertically polarized individual photon when analyzed at all different angles has quantum mechanical probabilities that generate the $\cos^2(\vartheta)$ law for passage. His didactic process gives a hint what would happen if the single photon presented in a given state were not accompanied by many many other copies. Einstein's reasoning from completeness is strikingly incomplete itself; unlike Bell's treatment of EPR reasoning, his argument against associating a wavefunction with only one particle is hardly mathematical. Which raises the question[12]: "What would happen if Einstein were right?" What experimental result should one expect?

Berko's teaching film provides a possible answer: When he presents vertical photons at low intensity to the ϑ-passing analyzer the result is more like scatter-plot than a regular "Law of Malus" $\cos^2(\vartheta)$ curve. Here is a screen shot:

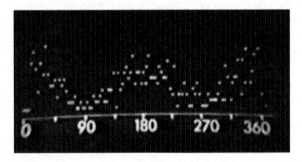

Fig. 7.1 Although the beam is actually an attenuated coherent state from laser source, the movie cites a quick calculation proving the chance of finding more than one photon at a time in the flight path from polarizer to analyzer is negligible. *Cf.* A.P.French & E.F.Taylor "Introduction to Quantum Physics," Chap. 6. Norton & Company (1978)

---

[11] Remarkably, its lab component is quite reminiscent of Mike Horne's independent innovation. At Stonehill College Mike set his physics students a semester long laboratory task, measuring all the fundamental constants of nature from Millikan oil-drop (now using latex spheres) for electron charge e, to the photoelectric effect for Planck's constant h, some speed of light experiments and Newton's gravity constant G via a Cavendish balance.

[12] A question strongly *emphasized* to me by Charles H. Bennett, in private communication (Conversation in Amherst MA, 2020).

There *is* an indication of a dip around 90° and 270°, but overall the fit is rather atrocious.

This is what a small ensemble of photons identically prepared in ket |V⟩ can do; what if there are NO repeats of the same exact state? We might expect each photon to have rather great statistical scatter, and perhaps no discernable pattern but true random scatter would emerge. So I imagined a new variation on Berko's old experiment.

Instead of always preparing the same Vertical state of polarization |V⟩ and rotating the analyzer, we keep the analyzer fixed at Vertical and rotate the polarizer so that, at most, only a single photon is prepared at a given ϑ during the entire run of thousands of counts. We also do a modern experiment with a down-conversion "polarization-entangled photon pair" source that provides heralded single photons, not the attenuated coherent state approximation used in the 1960's. If an individual lone particle does not have a wavefunction ψ, and *if* only ensembles of large numbers of identical photons obey the rules, we might get a scatter plot as we would for a Fig. 7.1 generated with even fewer photons counted.

On the other hand, if each particle has its own wavefunction |ϑ⟩, we once again will build up the statistics for a $\cos^2(ϑ)$ Law of Malus curve exactly as in the movie except ...

Except—if we bin the results correctly and vary the initial polarization sufficiently with each signal photon, we can do the experiment not only with never having two photons in the flight path simultaneously, but also without *ever* presenting more than one photon in the same exact polarization state |ϑ⟩. Then the only possible explanation for getting Malus' law is that the individual photons each must have had its own correct wavefunction.

Experimental Details and Its Theory

The "detailed description" section in past proposed experiments of mine[13] always turns out to be infinitely changeable (and ultimately changed) by those who actually know the experimental state of the art. But it is always worthwhile to go into the details. I propose we trigger an inflight polarization modification for the signal photon of a polarization entangled pair, once the idler has been detected. The detector is a polarization beam-splitter with two outputs, so as not to waste any photons.

---

[13] Some of my experimental proposals have been done, several helping to launch a sub-field. And every implementation has been different from my detailed description—quite changed, especially from the conference talks where I went into *deep* detail. They include 2π magnetic precession of neutrons, "Procrustean" entanglement concentration and SuperDense teleportation. *Cf.* my ideas in *PRL* **18**, 1102 (1967); *Phys. Rev.* **A53**, 2046 (1996); and *Quant. Inf. Proc.* **5**, #6, Springer Verlag, (XII 2006) with *PRL* **35**, 1053 (1975) & *PRL* **37**, 238 (1976) & *Phys. Lett.* **A54**, 425 (1975); *Nature* **409**, 1014 (2001); and *OSA Tech. Dig.* Online paper W6.44 (2013) https://doi.org/10.1364/QIM. 2013.W6.44.

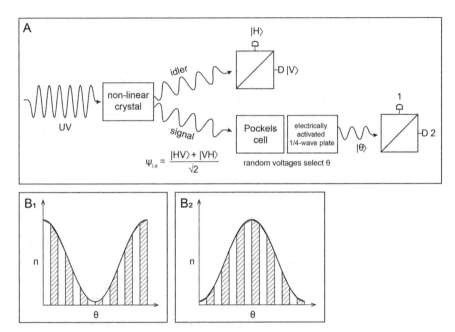

Fig. 7.2 The setup is in box A; its detectors 1 and 2 will show the curves depicted in boxes B_1 and B_2 when their counts are triggered by detector $|H\rangle$ and collected into bins of nearby angle settings. This illustration uses approximately 25.5° bins [$\pi/7$ radians]; it was rendered by Usha F Lingappa, recent CalTech Ph.D. See text for more explanation

I'm assuming Pockels cells or Kerr effect devices can be switched quickly enough and that we delay the signal photon in a long fiber or optical trombone suitably arranged to give those devices time to set. Thus the detector has a definite polarization photon to look at. Perhaps a worthy complexification is to provide random voltages to the polarizing devices, so no unsuspected hidden-variable kind of communication between the preparer and detector might influence the eventual compilation of a smooth probability of passage-&-detection curve. Here is a diagram of how to arrange things:

Box A of Fig. 7.2 above shows the setup. Squares with inscribed diagonals are polarizing beam-splitters. When the non-linear crystal is pumped with a UV laser, phase-matched polarization-entangled photon pairs are produced. Idler and signal beams enter different paths; detecting the polarization of the idler triggers a series of events that prepare and detect the signal. Assume the idler is detected as $|H\rangle$. Random voltages stimulate an electrically activated set of quarter wave plates to convert the signal into a $|\vartheta\rangle$. No two idler counts are allowed to cause the exact same voltage settings, so the statistics of identically prepared photons does not apply, contrary to the statistical interpretation: there **is** no ensemble; there aren't even any two identically prepared particles. Each state appears at most once. Counts of the signal photon are then recorded at detectors 1 and 2; binning those counts in nearby angles after a complete run produces the curves in Fig. 7.2B$_1$ [cos$^2(\vartheta)$] and 7.2B$_2$ [sin$^2(\vartheta)$], respectively. If the idler photon triggers the $|V\rangle$ output of its detector instead, the signal is a horizontal photon and the cos^2 & sin^2 curves are switched. That is, the curve depicted in B$_1$ will be associated with detector 2 and B$_2$ with detector 1.

By using both outputs of the idler detector, we can trace both complementary curves, cos$^2(\vartheta)$ and sin$^2(\vartheta)$ which ensue for the orthogonal polarization states created by the electrically operated variable orientation quarter-wave plates used to turn a Vertical polarization into an arbitrary linear polarization.[14]

Here's the theory of what happens. The UV-pumped down-conversion source produces an entangled pair

$$|S\rangle = \{|HV\rangle + |VH\rangle\}/\sqrt{2} \qquad (7.1)$$

After detecting the idler photon triggering the event and passing the electrically controlled elements, the signal photon has two different ψ states.

$$\begin{aligned}|\psi\rangle &= |\vartheta\rangle & \text{if}|idler\rangle \text{ was } |H\rangle \\ |\psi\rangle &= |(\vartheta + \pi/2)\rangle & \text{if}|idler\rangle \text{ was } |V\rangle\end{aligned} \qquad (7.2)$$

where $|\vartheta\rangle$ is defined as

$$|\vartheta\rangle = \begin{pmatrix} \cos\vartheta \\ \sin\vartheta \end{pmatrix}, \text{ and } |V\rangle \equiv \begin{pmatrix} 1 \\ 0 \end{pmatrix}, |H\rangle \equiv \begin{pmatrix} 0 \\ 1 \end{pmatrix} \qquad (7.3)$$

This implies B$_1$ counts roughly proportional to cos$^2(\vartheta)$ and B$_2$ to sin$^2(\vartheta)$; modified by the effective dark count noise signals which reduce the perfect contrast of these formulae, as shown in the boxes Fig. 7.2B$_1$ & B$_2$. In short, doing the experiment as described means that when the idler is detected, it converts the pure state of Eq. (7.1) into a mixed state with density matrix $\rho =$

$$= \begin{pmatrix} \cos^2\vartheta & \cos\vartheta * \sin\vartheta & 0 & 0 \\ \sin\vartheta * \cos\vartheta & \sin^2\vartheta & 0 & 0 \\ 0 & 0 & \sin^2\vartheta & -\sin\vartheta * \cos\vartheta \\ 0 & 0 & -\cos\vartheta * \sin\vartheta & \cos^2\vartheta \end{pmatrix}/2 \qquad (7.4)$$

---

[14] We could do even more, transforming $|V\rangle$ to an arbitrary point on the Poincaré sphere, while $|H\rangle$ goes to an orthogonal state. This was the trick used to prove fourfold success in teleporting polarization on a beam-entangled pair: the Harald Weinfurter-supervised experiment for Markus Michler's Ph.D., namely, showing all four curves, cos$^2(\vartheta \pm \vartheta_0)$ & sin$^2(\vartheta \pm \vartheta_0)$. It taught me the power of Charlie's intervening on Alice's half of the EPR pair—a precursor idea to SuperDense teleportation, cf. Experimental 2-Particle Quantum Teleportation *via Remote State Preparation* EQEC'98, Glasgow, p. 99 (1999) IEEE.

Discussion, Experimental Refinements and a Few More Reminiscences

Interest waxes and wanes in foundations and few of us can spend an entire career dealing with questions most physicists take for granted. But recent work has raised the interest in interpretation including a renewal of looking at Einstein's last stance.[15] And an overarching question like "Why the Quantum?" leads to all kinds of technical and detailed applications. In our collaboration's case it prepared us for a major shift in outlook from second and third generation quantum theory to the new field of quantum information (QI). Quantum states and their remarkable superposition and entanglement properties became tokens in QI, symbols in a special kind of message- & meaning-conveying alphabet.

Rather than yield "complementarity"-preserving, correspondence principle- obeying details about physical parameters of the particles, the quantum was deployed to do amazing, interesting things that ordinary computers, classical communication channels and Shannon information theory could not do. Instead of asking quantum states to tell about micro-reality one now asks what the existing micro-reality of states could do in the realms of computation, communication and information.

Which returns me to reminiscence of Mike Horne. Mike was a tremendously accomplished theorist; perhaps **the** greatest expert of his generation in straightfor- ward wave-function & Schrödinger QM, up to and including direct-product complex- vector tensor algebra. He was especially adept with the entangled states of all sorts of particles. He could do in a few error-free lines what field theorists might take many pages of creation and annihilation operators to calculate. And, since **his** Feynman- amplitude tracing was clear, it resulted in much more understanding. His modesty was legendary, perhaps best exemplified by a saying he applied to his own contribu- tions, a *ho-hummer*. Meaning perhaps one should react to the proposal by saying "ho HUM." "THIS is a ho-hummer," was a phrase I heard repeatedly when we initially met just a few weeks before the first international conference on Neutron Interferometry in Grenoble France, where he presented the neutron Fizeau effect. He was a master of reinterpretation as when he demonstrated the presence of a number filter, selecting bi-Photons, in one arm of the interferometer he showed Jeff Kimble's famous "light squeezing" experiment[16] to be. In another unpublished example, Mike led us to find an alternative understanding[17] of Alain Aspect's experiment showing a photon emitted from a single atom at two different places.

---

[15] "Einstein's Unfinished Revolution" Lee Smolin, Penguin Press (2019). See also APS news May 2018, *The Back Page*, & works by Flavio Del Santo of Austrian Academy of Sciences—which Zeilinger heads—and Del Santo's award-winning essay at https://www.aps.org/units/fhp/essay/upl oad/DelSantoEssay.pdf.

[16] **Squeezed states of light from an optical parametric oscillator,** Ling-An Wu, Min Xiao, & H. J. Kimble. JOSAB **4, Issue 10**, pp. 1465–1475 (1987) https://doi.org/10.1364/JOSAB.4.001465.

[17] Horne, MA Bernstein, HJ & Zeilinger, A "Interference Effect on a Photon Emitted by an Atom Localized in Two Separated Regions" (unpublished PRL comment article, 1985).

The Dotted Line of QM: An Essential Refinement

The experimental diagram in Fig. 7.2A raises a wonderful little point of re-interpretation that makes me wish Mike were here to see and turn over with me. It relates to a certain *dotted line* dividing the quantum mechanical realm from the classical reality of our everyday world. And it emphasizes a subtle, lightly mentioned, point about the setup in Fig. 7.2—the need for a delay to provide time to activate the devices making the polarization transformation. They are far from instantaneous.

In our graduate school quantum courses we heard about a "dotted line." It emphasized a relation of the theory to classical physics; a relationship connected to Bohr's correspondence principle. To get answers from a quantum system some measurement or definitive evolution has to happen. So there is always a dotted line separating that part of the world being described by QM from that which obeys classical physics. The line was said to be quite movable, perhaps between Wigner's friend and a meter, or between Wigner and his friend—maybe, even, as my great UCSD Professor Meir Weger remarked, between the publishing physicist and the PRL office. (Asher Peres once claimed[18] the division was a matter of how much money one could spend in order to measure the large observing apparatus in quantum detail).

Note that there is a different kind of dotted line implicit in Fig. 7.2A, which bears strongly on my experiment's interpretation. Implicit in my discussion of the setup and Eq. (7.4), I am seeing a vertical dotted line between preparation and measurement drawn right after the "electrically activated quarter-wave plates." Everything to the left is considered preparation as triggered by detection of an idler photon. The wiggly line at the end of the bottom beam has been made to be a single photon in the state $|\vartheta\rangle$, the very essence of my proposal.

But one could draw a dotted line earlier in the lower beam just before the signal photon enters the Pockels cell, and consider all the apparatus to its right as part of the detector; in which case the experiment is making a definite polarization photon (labelled *signal* in Fig. 7.2A), whose $|H\rangle$ or $|V\rangle$ character can be determined by correlation with the $|V\rangle$ or $|H\rangle$ output of the *idler* detectors. And the larger amount of equipment to the right of the dotted line is then, in effect, using random voltages to change the detector orientation ϑ, from being fixed at $\vartheta = 0$, to every different value—sampling none more than once of course.

In all of this there's a key factor that I have left out: The delay is in the *signal* line. The signal photon has to be delayed until the electrically operated devices are correct, which is why the state in Eq. (7.4) is a mixture. But we could put the delay in the idler beam, making the measurement only after the preparation apparatus has been switched randomly to its new setting. Then measuring the idler polarization simply chooses the moment when we actually have found an emitted pair. The state has remained entangled and is pure when both photons' polarizations are measured. This is a better way to do the experiment. Its theory is modified from what we had above, as follows.

---

[18] *Am J Phys.* **42** (1974) 886. See also his **Quantum Theory** Kluwer Academic (1994), p. 376.

Experiment with *Idler* Delay Line

The setup is again as in Fig. 7.2, but there is a delay in the idler beam [still not shown]. While the UV pump creates a down-conversion pair, random voltages electrically activate a set of quarter wave plates that convert it by rotating the linear polarization through an angle ϑ, *i.e.* transform it by R(ϑ). Where

$$R(\vartheta) = \begin{pmatrix} \cos \vartheta & -\sin \vartheta \\ \sin \vartheta & \cos \vartheta \end{pmatrix} \tag{7.5}$$

this transforms the state in Eq. (7.1), $|S\rangle = (|H\rangle \otimes |V\rangle + |V\rangle \otimes |H\rangle)/\sqrt{2}$ into

$$|S'\rangle = (|H\rangle \otimes |\vartheta\rangle + |V\rangle \otimes |(\vartheta + \pi/2)\rangle)/\sqrt{2} \tag{7.6}$$

[where I have made the outer-product and entangled character of our states a bit more obvious by using the $\otimes$ symbol for Cartesian product]. When an idler is successfully detected in the upper beam, it triggers a quick check—that the Pockels cell and quarter wave plates were indeed already set, not just turning on or switching to a new voltage setting—then activates the final detectors of signal polarization. Detector 1 and 2 counting curves are still exactly as cited in the figure and discussion above, but with this implementation there is no "dotted line" ambiguity. Everything is necessarily quantum mechanical and coherent until the final (joint) measurement of the polarization of the idler & signal photons. The dotted line *must* be drawn after the voltage-dependent polarization transformations.

Even without this little logical twist one could say I am proposing a real *ho-hummer* of an experiment. Which brings me right back to my first meeting with Mike, during which visit he repeatedly announced[19] his "neutron Fizeau effect" idea was a "ho-hummer." In the mid-nineteenth century, Hippolyte Fizeau showed that introducing a moving medium to a *light* interferometer induced a velocity-dependent phase shift. Mike Horne analyzed what would happen in a matter-wave interferometer. In the last few weeks, now 42 years later, I realized he may simply have been referring to the absence of phase shift for *neutron* radiation when the motion of refractive medium was entirely within boundaries fixed in space, no matter what the velocity of the moving medium.

One further refinement is potentially worthy of mention. It is hard to characterize how completely this experiment, when successful, violates the implication of Einstein's notion, since the idea that ψ is not applicable to a single photon has no definite mathematical form. But if one converted each predicted quantum outcome to a binary yes-or-no of agreement with QM, the possibility of Einstein's suggestion

---

[19] The humility was truly impressive to a brash former high-energy theorist like me, "full of self" and pride for proposing the first practical demonstration that a 2π-radian rotation of fermions can be detected in **neutron** interferometry. It completely switches constructive to destructive interference and vice versa; and 4π rotation restores the original pattern.

being correct **and** producing a DIFFERENT answer can be quantified. Each correct answer in agreement with QM has only a 50% chance on Einstein's most radical possibility: that a single representative of a state knows nothing of the probabilities given—for a large ensemble—by ψ. The setup for such an extended experiment is quite curious, and raises additional philosophical questions. Sketch it for yourself and you may find interesting details, as I'm sure Mike Horne would have! But this chapter is already long enough, and AGAIN, Einstein did not explicitly say the results are fully random[20] if you have only one representative particle. So we shall leave the enhanced binary-choice experiment "outside the scope of this paper," and end here.

Acknowledgments Thanks to Mike Horne and our colleagues for the excellent work & play which has contributed so much to the world, the field(s) of physics and the development of 20th & 21st century quantum sciences. Prominent among them, of course are Daniel Greenberger, Clifford Shull, Anton Zeilinger and their people: students, collaborators and families. Thanks for the wonderful collaborative work and the honour of serving as Principal Investigator on all those groundbreaking experiments and unbelievable theoretical explorations of the question "Why the Quantum?" They have enriched my own career and growth immensely. Thanks to Mike Fortun, the Institute for Science & Interdisciplinary Studies (IS&IS) and all its supporters, co-Workers & volunteers and all our collaborators as well. Thanks to Hampshire College whose unique freedom, faculty support policies and innovative mission have always provided a playground for independent thought and creativity, especially among my peers in the school of Natural Sciences & in the other two science-named divisions, social and cognitive. Thanks to Paul G Kwiat, whom I originally met through the NSF collaboration on Quantum Interferometry, with whose people & research group several implementations of my experimental suggestions have been brought to fruition. Special thanks to Kenneth Finkelstein and Charles H Bennett, Theodor Hänsch, Harald Weinfurter and Jim Gates for conversations and other interactions, and to my former students Usha F. Lingappa (for the illustration and for the discussions of 2-state quantum mechanics from the first undergraduate year on) and Lee Smolin (for looking at a draft of this article). Of course my wife Mary and our children, who have always given me the leeway and support needed to pursue interests in physics and beyond, have my deepest undying love and appreciation.

---

[20] Certainly those points where the interference pattern has zeroes (nodes) and places with anti-nodal, 100%, probabilities should exist as definite NO and YES predictions even for one particle at a time.

Chapter 8
Complementarity and Decoherence

Don Howard

Introduction

Ninety-five years after the advent of modern quantum mechanics there is still no consensus on the question of how a seemingly definite macroworld emerges from an indefinite microworld, the puzzle being especially acute in the context of the measurement problem. Linear Schrödinger dynamics of necessity evolves the pre-measurement joint state of the observed object and the measuring instrument, a product of instrument and object eigenstates, into an entangled, pure, joint state that is a superposition over joint, instrument-object eigenstates. And yet, when we look, we seem always to find the instrument in a definite, instrument eigenstate. Standard quantum mechanics explains this, if one can really call it an explanation, by the ad hoc postulation of a second mode of state evolution exclusively for measurements, what John von Neumann dubbed "Process 1" (von Neumann 1932, 186), which evolves the pre-measurement joint instrument-object state into a mixture over instrument-object eigenstates, some one of which then magically and non-deterministically manifests itself in what was termed the reduction or the collapse of the wave packet. When combined with the Born rule for interpreting quantum mechanical probabilities, the trick works well enough to satisfy the needs of the work-a-day physicist. But almost no one believes that it is true. After all, a measurement is just a physical interaction and there is every reason to believe in the universal truth of linear Schrödinger dynamics. So why should there be a dramatically different dynamics for that subset of interactions that we, for practical reasons, deem "measurements." Nature cannot care about whether some physical interactions are put to special use by these historical accidents called "human beings." Our having happened to evolve cannot have made a difference in the fundamental laws of the universe.

D. Howard (✉)
Department of Philosophy and History and Philosophy of Science Graduate Program, University of Notre Dame, Notre Dame, USA
e-mail: dhoward1@nd.edu

© Springer Nature Switzerland AG 2021   151
G. Jaeger et al. (eds.), *Quantum Arrangements*, Fundamental Theories of Physics 203,
https://doi.org/10.1007/978-3-030-77367-0_8

Numerous suggestions emerged over the years for how we might solve this puzzle. David Bohm and his low church epigones assert that an in-principle epistemically inaccessible, sub-quantum physics—hidden variable theory—has fixed everything to determinate trajectories and, derivative from that, determinate possessed values of all observables, all the while hiding the weirdness that does all of the quantum physics heavy lifting, entanglement, in that ephemeral *j'ne sais quois* that they call the "quantum potential," which, thanks to its being labeled a "potential," is denied, by implication, the status of physical reality (Bohm 1952a, b). Hugh Everett and his high church fellow believers, who all, on principle, disdain Ockhamist, common sense ontological parsimony, tell us that, upon that uniquely, subjectively, human act of observation, the universe magically divides into n daughter universes (where n can range from 2 to ∞) in each of which one term, one eigenstate, in the pure, joint, entangled, instrument-object, post-measurement, superposition state is realized, this splitting of the entire universe ramifying, micro-moment by micro-moment, since and until eternity every time anyone anywhere blinks, even though the observer has no cognitive access to the splitting or those other universes, being aware only of what he or she perceives with the measuring apparatus (Everett 1957). Other responses to the measurement problem aspire to be less metaphysical and more physical. Giancarlo Ghirardi, Alberto Rimini, and Tullio Weber bite the bullet and just posit that collapse is a stochastic process, abstaining from a guess as to how and why it happens (Ghirardi et al. 1985). And, long before Roger Penrose claimed the idea as his own and made it famous (Penrose 1986), the Hungarian physicist, Frigyes Károlyházy, suggested, cleverly, that collapse might be gravitationally induced (Károlyházy 1966). Most amusing of all was the suggestion by my *Doktorgrossvater*, Eugene Wigner, that wave packet collapse was induced by the subjective consciousness of the observer (Wigner 1962), which is, no doubt, why my *Doktorvater*, Wigner's student, Abner Shimony, declared his presence on the stage in the early 1960s with his classic paper on "The Role of the Observer in Quantum Theory," which provided a masterful review of the measurement problem since the early work of von Neumann and Fritz London and Edmond Bauer (London and Bauer 1939) and included trenchant critiques of both Everett and the idea of mind-induced wave packet collapse (Shimony 1963).

One would like to think that even the most ardent supporters of these alternatives to the standard quantum mechanical treatment of the measurement problem would grant that they are each, in their own way, as ad hoc as von Neumann's original approach. None of them can be said to be based on physically well-motivated principles, although modern Everettians will insist that their program is well motivated, since it is, they say, just orthodox quantum mechanics without the collapse postulate, and that, by the way, that many-worlds business is not an absurd, metaphysical extravagance, it is just the way the world must be if quantum mechanics without collapse is true (Wallace 2012). We are given no causal account of how mind or gravity induce collapse, just the assumption that they do. And Bohmian mechanics is just a mathematical sleight-of-hand, because, as mentioned, it just hides entanglement in the quantum potential and hopes that no one will notice.

What none of these alternative programs or standard quantum mechanics have ever questioned is the crucial premise that, when we measure, when we observe, we always find the observed system to possess a definite value of the parameter of interest, be it position, momentum, spin, polarization, or what have you. That is the assumption that drives everyone from von Neumann to Everett to one or another, desperate and dubious expedient. It might seem obvious that measurements always yield definite results. But what if that were not so? And notice that, absent any compelling physical basis for distinguishing measurement interactions from all others, it cannot be so, because the otherwise universally valid, linear Schrödinger dynamics makes it impossible. What if, perhaps as a consequence of the scale on which we percipient humans exist, we have been living under an illusion of definiteness? What if the world were really always and everywhere indefinite, as quantum mechanics without collapse says it must be, but that it is so in such a way that we do not experience it that way? It was that question that gave birth many years ago to the program that is known by the unfortunately very confusing name of "decoherence theory."

That name, "decoherence theory" is confusing because it wrongly suggests that it is all about wave packet collapse, and scads of authors persist in using the term "decoherence" as meaning exactly that. Ironically, "decoherence" theory posits the opposite, that there is no collapse but that Schrödinger dynanmics applied to the interaction between a measured object and the trillions of environmental degrees of freedom in which it is bathed drives that joint system into a state indistinguishable from a mixture over definite states.

Modern decoherence theory originated in the work of Heinz Dieter Zeh in the 1970s and Wojciech Zurek in the 1980s (Zeh 1970, 1973; Kübler and Zeh 1973; Joos and Zeh 1985; Zurek 1981, 1982). But already from the middle 1940s through the early 1960s, Werner Heisenberg and others—Leon Rosenfeld, Aage Petersen, Hilbrand Groenwald, Carl Friedrich von Weizsäcker, and Günther Ludwig—worked explicitly on Niels Bohr's idea that the dynamics of the measurement interaction, that is, the coupling of degrees of freedom of the observed system to instrument and environmental degrees of freedom, was responsible for the appearance of classicality in such forms as the seeming irreversibility of measurement outcomes (see Schlosshauer and Camilleri 2017).

Stated carefully, the central idea of decoherence theory, properly understood, is this. When a system or a system-instrument pair is entangled with a large number of environmental degrees of freedom, the purely quantum dynamics of the interaction with the environment, meaning ordinary, linear, quantum mechanical, Schrödinger dynamics, drives the system-instrument-environment complex into a pure state that is observationally indistinguishable from a mixture over eigenstates. In this way, quantum mechanics, itself, explains the emergence of a simulacrum of classicality, even though, in principle, the world at all scales remains purely quantum. Morever, it is typical of decoherence that the dynamics of the system-instrument-environment interaction picks out a privileged basis, what is generically termed a "pointer basis" or an "einselected basis," and in many cases, but not always, that basis will be one close enough to a position basis as to be functionally equivalent to it for all practical purposes. Such a basis is privileged in the sense that it is robust with respect to perturbations.[1]

But there remains a puzzle. In what precise sense is the pure state into which the decoherence dynamics drives the system-instrument-environment complex "observationally indistinguishable" from a mixture over eigenstates? What kind of a mixture over eigenstates are we talking about? It cannot be the improper mixture that one obtains by tracing over all instrument plus environment degrees of freedom, because that mixture is easily distinguishable from the pure state. So, again, what kind of mixture is it that is observationally indistinguishable from the pure state? It might come as a surprise to learn that, in order to answer this question, we must take a step back in time to reconsider the central interpretive proposal of the very same Bohr who first suggested that it was the coupling of object and environment degrees of freedom that gave rise to the appearance of classicality, the doctrine of complementarity.

Complementarity, Entanglement, and Classical Concepts

There is widespread misunderstanding of Bohr's doctrine of complementarity, much of it owing not to Bohr's alleged obscurity, but to a failure to understand the extent to which and manner in which complementarity was seen by Bohr to be a direct logical consequence of the physical fact of entanglement. Your first thought might now be that it is impossible for Bohr to have regarded complementarity as a consequence of entanglement, because Bohr introduced complementarity in his 1927 "Como" lecture, "The Quantum Postulate and the Recent Development of Atomic Theory" (Bohr 1928), whereas entanglement was only introduced in the mid-1930s by Erwin Schrödinger in a series of three papers prompted by the Einstein, Podolsky, and Rosen (EPR) paper and Bohr's reply to that (Schrödinger 1935a, 1936, b; Einstein et al. 1935; Bohr 1935). Schrödinger there describes entanglement in these words:

> If two separated bodies, about which, individually, we have maximal knowledge, come into a situation in which they influence one another and then again separate themselves, then there regularly arises that which I just called *entanglement* [*Verschränkung*] of our knowledge of the two bodies. At the outset, the joint catalogue of expectations consists of a logical sum of the individual catalogues; during the process the joint catalogue develops necessarily according to the known law [linear Schrödinger evolution] ... Our knowledge remains maximal, but at the end, if the bodies have again separated themselves, that knowledge does not again decompose into a logical sum of knowledge of the individual bodies. (Schrödinger 1935a, 827)

But, if you thought that, you would be wrong, because by 1927 nearly everyone understood that entanglement was an ineluctable feature of quantum mechanics. What Schrödinger did in 1935 was merely to give a helpful new name to and carefully explicate what had long been common knowledge.

---

[1] There remains no better introduction to decoherence theory than Schellhammer 2007, though I also strongly recommend Crull (2011), Chap. 2, especially for the non-physicist.

Entanglement in the Late 1920s and Early 1930s

The most striking proof of the commonplace understanding of the role of entanglement in quantum mechanics by 1927 is surely the fact that Albert Einstein's own failed attempt at a hidden variables interpretation of quantum mechanics in the Spring of 1927 ran aground on precisely these shoals. The story, in brief, is this. In early May of 1927 Einstein presented a paper to the Berlin Academy with the title, "Does Schrödinger's Wave Mechanics Determine the Motion of a System Completely or Only in the Statistical Sense?" (EA 2–100), in which he developed a hidden variables interpretation that he described in a letter to Paul Ehrenfest of May 5, 1927 in these words: "I have also now carried out a little investigation concerning the Schrödinger business, in which I show that, in a completely unambiguous way, one can associate definite movements with the solutions, something which makes any statistical interpretation unnecessay" (Einstein to Ehrenfest, May 5, 1927, EA 10–162). Only after the lecture was set in proofs for publication in the proceedings of the Berlin Academy did Einstein and his then assistant, Jakob Grommer, notice a problem, which was set out in a "Note Added in Proof":

> I have found that the schema does not satisfy a general requirement that must be imposed on a general law of motion for systems.
>
> Consider, in particular, a system $\sum$ that consists of two energetically independent subsystems, $\sum_1$ and $\sum_2$; this means that the potential energy as well as the kinetic energy is additively composed of two parts, the first of which contains quantities referring only to $\sum_1$, the second quantities referring only to $\sum_2$. It is then well known that
>
> $$\Psi = \Psi_1 \cdot \Psi_2$$
>
> where Ψ_1 depends only on the coordinates of $\sum_1$, Ψ_2 only on the coordinates of $\sum_2$. In this case we must demand that the motions of the composite system be combinations of possible motions of the subsystems.

The indicated scheme [Einstein's own hidden variables model] does not satisfy this requirement. In particular, let μ be an index belonging to a coordinate of $\sum_1$, ν an index belonging to a coordinate of $\sum_2$. Then $\Psi_{\mu\nu}$ does not vanish.

What Einstein had discovered, to his dismay, was that the very entanglement that so bothered him about Schrödinger wave mechanics was lurking in his own hidden variables interpretation. The paper was never published.[2]

In fact, Einstein had been brooding for a very long time about the failure of mutual independence of spatially separated, quantum-level systems.[3] This was a major focus of puzzlement for him in his work on Bose–Einstein statistics in 1924 and 1925, as when he wrote in the second of his three papers on the topic:

> Bose's theory of radiation and my analogous theory of ideal gases have been reproved by Mr. Ehrenfest and other colleagues because in these theories the quanta or molecules are

---

[2] For a detailed discussion of the manuscript, see Belousek (1996).

[3] I discuss this history in detail in Howard (1990).

not treated as structures statistically independent of one another, without this circumstance being especially pointed out in our papers. This is entirely correct. If one treats the quanta as being statistically independent of one another in their localization, then one obtains the Wien radiation law; if one treats the gas molecules analogously, then one obtains the classical equation of state for ideal gases, even if one otherwise proceeds exactly as Bose and I have ... It is easy to see that, according to this way of calculating [Bose-Einstein statistics], the distribution of molecules among the cells is not treated as a statistically independent one. This is connected with the fact that the cases that are here called "complexions" would not be regarded as cases of equal probability according to the hypothesis of the independent distribution of the individual molecules among the cells. Assigning different probability to these "complexions" would not then give the entropy correctly in the case of an actual statistical independence of the molecules. Thus, the formula [for the entropy] indirectly expresses a certain hypothesis about a mutual influence of the molecules–for the time being of a quite mysterious kind–which determines precisely the equal statistical probability of the cases here defined as "complexions." (Einstein 1925, 5-6)

Schrödinger, for one, was so confused by this that Einstein had to explain:

In the Bose statistics employed by me, the quanta or molecules are not treated as being independent of one another .... A complexion is characterized through giving the number of molecules that are present in each individual cell. The number of the complexions so defined should determine the entropy. According to this procedure, the molecules do not appear as being localized independently of one another, but rather they have a preference to sit together with another molecule in the same cell. One can easily picture this in the case of small numbers. [In particular] 2 quanta, 2 cells:

Bose-statistics				independent molecules		
	1st cell	2nd cell			1st cell	2nd cell
1st case	●●	—		1st case	I II	—
				2nd case	I	II
2nd case	●	●		3rd case	II	I
3rd case	—	●●		4th case	—	I II

According to Bose the molecules stack together relatively more often than according to the hypothesis of the statistical independence of the molecules. (Einstein to Erwin Schrödinger, 28 February 1925, EA 22-002).

In a PS to this letter, Einstein adds that the difference between Bosonic and Boltzmann statistics would only be significant for dense gases and that "there the interaction between the molecules makes itself felt, the interaction which, for the present, is accounted for statistically, but whose physical nature remains veiled."

As I have explained in considerable detail elsewhere (Howard 1990), Einstein's worries about the failure of mutual independence in the quantum realm go all of the way back to his 1905 paper on the photon hypothesis (Einstein 1905), in which the postulate that electromagnetic energy lives in the form of corpuscle-like, mutually independent light quanta is valid only in the high-energy, Wien regime, implying, as Einstein well understood, that outside of the Wien limit, which is to say, for all intents and purposes, in all real situations, that model of mutually independent light quanta would fail. He explained this in a letter to Hendrik Lorentz on 23 May, 1909:

> I must have expressed myself unclearly in regard to the light quanta. That is to say, I am not at all of the opinion that one should think of light as being composed of mutually independent quanta localized in relatively small spaces. This would be the most convenient explanation of the Wien end of the radiation formula. But already the division of a light ray at the surface of refractive media absolutely prohibits this view. A light ray divides, but a light quantum indeed cannot divide without change of frequency …
>
> As I already said, in my opinion one should not think about constructing light out of discrete, mutually independent points. I imagine the situation somewhat as follows: … I conceive of the light quantum as a point that is surrounded by a greatly extended vector field, that somehow diminishes with distance. Whether or not when several light quanta are present with mutually overlapping fields one must imagine a simple superposition of the vector fields, that I cannot say. In any case, for the determination of events, one must have equations of motion for the singular points in addition to the differential equations for the vector field. (Einstein to Lorentz, May 23, 1909, CPAE, Doc. 163, 193)

This is the birth of wave-particle duality and the concept of the guiding field, which latter is postulated precisely to explain the failure of mutual independence of the particles thanks to interference effects involving the wave fields. Of course were the overlapping fields a "simple superposition," that would not yield entanglement, but notice that Einstein remarks that he "cannot say" whether it would be so.

Where, elsewhere in the literature, do we find evidence that entanglement was well understood to be a fundamental feature of quantum mechanics before Schrödinger coins the term in 1935? Consider two of the many discussions of the issue. The first is from Hermann Weyl's *Gruppentheorie und Quantenmechanik* [*Group Theory and Quantum Mechanics*], where, in section ten of chapter two, of the 1931s edition, "Mehrkörperproblem. Zusammensetzung." ["Problem of Many Bodies. Composition."], Weyl writes:

> *Conditions that will not allow a further increase in homogeneity within c* [a composite system] *nonetheless need not require a maximum in this respect within the partial system* **a**. *Furthermore: If the state of* **a** *and of* **b** *are known, the state of* **c** *is still not uniquely specified.* For a positive definite Hermitian form $\|a_{i,k,\ i'k'}\|$ in the total space, which describes a statistical aggregate of states **c**, is not uniquely determined by the Hermitian forms that spring from it

$$\sum_k a_{ik,i'k}, \quad \sum_i a_{ik,ik'}$$

> in the spaces R, S. In this concise form the assertion that "*the whole is greater than the sum of its parts*," which has recently been elevated to an article of faith by the Vitalists and the Gestalt Psychologists, holds good in the quantum theory. (Weyl 1931, 92)

The second example comes form Wolfgang Pauli's masterful article on "Die allgemeinen Prinzipien der Wellenmechanik" ["The General Principles of Wave Mechanics"] in the1933 second edition of the *Handbuch der Physik* [*Handbook of Physics*]. Though technically a handbook article, this is actually a small book, at 149 pages, and it functioned as an introductory textbook for a couple of generations of physics students, meaning that entanglement was by this time a well established part of the textbook literature on quantum mechanics (see Howard 2009). Section 5, "Wechselwirkung mehrerer Teilchen. Operatorkalkül." ["Interaction of Several Particles. Operator Calculus."], opens with these words:

> The manner in which a composite system consisting of several subsystems is described in the quantum theory is of fundamental importance for the theory and its most distinguishing characteristic. It shows, on the one hand, the fruitfulness of Schrödinger's idea of introducing a ψ-function that satisfies a linear equation and, on the other hand, the purely symbolic character of this function, which differs from the wave functions of the classical theory . . . in a fundamental way.

> If a we have a system of several particles, one obtains no satisfactory description of the system by the specification of the probability for one of the particles to be found at a specific location. (Pauli 1933, 111)

Pauli introduces probabilities for the N separate particles:

$$W_1(q_1, q_2, q_3), W_2(q_4, q_5, q_6), \ldots, W_N(q_{3N-3}, q_{3N-1}, q_{3N})$$

He then remarks about the relationship between these separate probabilities and the joint probability, $W(q_1 \ldots q_f)$, where $f = 3\,N$:

> Only in a special case is the knowledge of the functions $W_1 \ldots W_N$ equivalent to a knowledge of the function $W(q_1 q_2 \ldots q_f)$, namely, if this function W decomposes as a product:

$$W(q_1 \ldots q_f) = W_1(q_1, q_2, q_3), W_2(q_4, q_5, q_6), \ldots, W_N(q_{3N-3}, q_{3N-1}, q_{3N})$$

> In this specific case we say that the particles are statistically independent of one another. (Pauli 1933, 112)

To what does this correspond from the point of view of the Hamiltonian? Pauli explains:

> An additive decomposition of the Hamiltonian operator in independent summands thus corresponds to a product decomposition of the wave function in independent factors. This is in accord with the circumstance that, in the case of statistically independent particles, the probability $W(q_1 \ldots q_f; t)$ can be decomposed as a product. (Pauli 1933, 114).

We began this review of the early appreciation of entanglement's fundamental importance in quantum mechanics in order to make plausible the suggestion that Bohr's doctrine of complementarity was, in fact, driven precisely by his appreciation of the deep implications of entanglement for the interpretation of the quantum theory. Is there direct evidence of Bohr's also having shared in this common understanding?

Entanglement as the Basis of Complementarity

That Bohr, too, knew of the deep significance of quantum entanglement and that
it was the premise from which complementarity is derived is plainly evident from
Bohr's own words where he first introduces the doctrine of complementarity in the
Como paper:

> Now, the quantum postulate implies that any observation of atomic phenomena will involve
> an interaction with the agency of observation not to be neglected. *Accordingly, an independent
> reality in the ordinary physical sense can neither be ascribed to the phenomena nor to the
> agencies of observation.* [My italics.] ....
>
> This situation has far-reaching consequences. On one hand, the definition of the state of a
> physical system, as ordinarily understood, claims the elimination of all external disturbances.
> But in that case, according to the quantum postulate, any observation will be impossible, and,
> above all, the concepts of space and time lose their immediate sense. On the other hand, if
> in order to make observation possible we permit certain interactions with suitable agencies
> of measurement, not belonging to the system, an unambiguous definition of the state of
> the system is naturally no longer possible, and there can be no question of causality in the
> ordinary sense of the word. The very nature of the quantum theory thus forces us to regard
> the space-time co-ordination and the claim of causality, the union of which characterizes the
> classical theories, as complementary but exclusive features of the description, symbolizing
> the idealization of observation and definition respectively. (Bohr 1928, 580)

These few sentences have been, of course, widely quoted and just as widely
misunderstood. Start with the first sentence: "The quantum postulate implies that
any observation of atomic phenomena will involve an interaction with the agency
of observation not to be neglected." Many commentators have read this through the
lens of Heisenberg's advocacy of a disturbance analysis of measurement, according
to which the observer-observed interaction changes uncontrollably the state of the
observed object, as in Heisenberg's famous electron-microscope thought experiment, wherein the back-scattering of an electron hitting an atomic scale object yields
precise position information only on pain of an uncontrollable change in the object's
momentum (see, for example, Heisenberg 1930, 20–30). The irony is, of course, that
it is precisely Bohr's rejection of that way of explicating quantum indeterminacy
that generated the tension between Bohr and Heisenberg after the latter's submission
of his classic paper on indeterminacy to the *Zeitschrift für Physik* in 1927 (Heisenberg 1927). As Bohr explains in the Como paper, indeterminacy is more about the
"possibilities of definition" than about the possibilities of "observation" (Bohr 1928,
585–589), and Heisenberg's disturbance analysis of indeterminacy also wrongly
assumes the existence of a precise, pre-measurement value of the momentum, for,
otherwise, there would be nothing to be disturbed.

Consider, next, the italicized words: "*Accordingly, an independent reality in the
ordinary physical sense can neither be ascribed to the phenomena nor to the agencies
of observation.*" They are often held up as the definitive evidence for the claim that
Bohr's complementarity interpretation of quantum mechanics is a species of anti-
realism (see, for example, Baggott 2011, 111). But Bohr does not say here that no
reality can be attributed to the phenomenon or to the agencies of observation. What

he says is that they can be ascribed no "independent reality," and he says this not only about the quantum-scale object of measurement but also and equally about the macroscopic measurement apparatus. If one were to read this sentence as a denial of the reality of the atomic world then, by parity of reasoning, Bohr would be committed to denying the reality of the macroworld, which, surely, he did not intend.

What, then, did Bohr mean? Once one understands that an appreciation of the deep significance of entanglement in the quantum world was commonplace by 1927, one sees that Bohr's meaning is plain: A measurement requires a physical interaction between object and instrument, from which it follows that the post-measurement, joint, object-instrument state is, necessarily, an entangled one. It is as simple as that. The object and instrument do not lack reality *totaliter*, rather they lack independent reality, the emphasis now being on the adjective, "independent," because they form an entangled pair. Thus, complementarity is, for Bohr, a direct consequence of quantum entanglement.

As an aside, note that Bohr is not speaking here about a relationship between the observer and the observed, but, instead about "the phenomena" and the "agencies of observation." This is characteristic of Bohr's writings on the interpretation question. He is always careful to speak not of a conscious, human observer but of the physical measuring apparatus. Contrary to the tradition from von Neumann, through London and Bauer, and on to Wigner, where it is the act of observation or even the registration of measurement result in the consciousness of the observer that induces wave packet collapse, Bohr accords no special role to the observer, or rather, to the observer in his or her subjective aspect. Instead, for Bohr, it is all about the quantum physics of the physical interaction between the object and the measuring apparatus, which may, of course, include the observer in his or her physical aspect. But that a human observer might be present, eyeballing the apparatus, is, for Bohr, incidental and inconsequential to the analysis of the consequences of entangling the object and the instrument in measurement. That so many think that Bohr does accord the subjective, human observer a special role (see, for example, Katsumori 2011, 62) is, as with the misreading of the line about "an interaction with the agency of observation not to be neglected," another instance of the all-too-common habit of reading or misreading Bohr's philosophy of physics through the lens of Heisenberg's philosophy of physics, a persistent error that has done much to confuse our understanding of Bohr and complementarity, as I have argued elsewhere (Howard 2004).

Entanglement entails complementarity. But what is complementarity? Bohr says that it is a relationship of reciprocal exclusion between "space–time co-ordination and the claim of causality." These words have occasioned as much confusion as have Bohr's comments about the object and instrument lacking independent reality. But to figure out what Bohr means it helps to recall the broadly Kantian philosophical tradition in which Bohr was educated, especially through the philosophy lectures in Copenhagen of the noted Harald Høffding, who was a close, personal friend of Bohr's father and a frequent guest in the Bohr household when Niels was a young boy (Moore 1966, 14–15, 406–407, 432). The relationship between Høffding and the younger Bohr was so close that Høffding recruited Niels's help with the preparation of the fifth edition of his logic textbook (Høffding 1907; Röseberg 1992, 32). Bohr was so enamored with philosophy that he contemplated writing a book on epistemology and enthusiastically joined other students of Høffding's in a lively, philosophical discussion group (Röseberg 1992, 29–30; Moore 1866, 17–18).

By the end of the nineteenth century, thinkers like Arthur Schopenhauer had pared down the Kantian a priori to just space and time as the necessary, a priori forms of outer and inner intuition and causality as an a priori category of the understanding, these a priori structures playing obviously compatible, conjoint roles in judgment, hence in cognition generally, including scientific cognition (Schopenhauer 1819). Bohr now is arguing that, as a consequence of entanglement, what were compatible structures of cognition in a world as described by classical physics must now be seen as incompatible but complementary aspects of scientific knowledge in a world as described by quantum mechanics. But what, more precisely, are "space–time coordination" and "the claims of causality" in the setting of quantum physics? Bohr, himself, makes that clear when, two pages later, he writes:

> In the language of the relativity theory, the content of the relations (2) [the Heisenberg position-momentum and energy-time indeterminacy relations] may be summarised in the statement that according to the quantum theory a general reciprocal relation exists between the maximum sharpness of definition of the spacetime and energy-momentum vectors associated with the individuals. This circumstance may be regarded as a simple symbolical expression for the complementary nature of the space-time description and the claims of causality. At the same time, however, the general character of this relation makes it possible to a certain extent to reconcile the conservation laws with the space-time coordination of observations, the idea of a coincidence of well-defined events in a space-time point being replaced by that of unsharply defined individuals within finite space-time regions. (Bohr 1928, 582)

"Space–time coordination," therefore, means simply the assignment of definite spatial and temporal locations to a system, and "the claims of causality" means energy and momentum conservation. There is nothing the least bit obscure in this. Putting it all together, Bohr is saying that we cannot ascribe definite positions and times without performing measurements, which requires a physical interaction between the object and the instrument, but, since the object will then no longer have a well-defined, independent state, thanks to entanglement, there can be no talk of energy and momentum conservation, not because energy and momentum conservation are violated, but because they can no longer be clearly formulated, as least not in the classical sense, since that requires the ascription of well-defined dynamical states.

Note that Bohr's saying here that the Heisenberg indeterminacy relations are a "simple symbolical expression" for complementarity puts the lie to another, widespread, false claim about Bohr's doctrine of complementarity, which is that Bohr changed his mind about complementarity after the EPR paper, replacing the allegedly confused, 1927 version of complementarity with complementarity as a relationship between observables represented by non-commuting operators, after realizing, under the press of EPR, that complementarity was not only a relationship between a microsystem and a macroscopic measuring apparatus, but that it obtained also between the properties of two interacting but then separated quantum scale objects (see, for example, Beller and Fine 1994, 10–16). As it turns out, that had been Bohr's understanding of complementarity from the very beginning. Complementarity in the setting of the relationship between a macroscopic apparatus and a quantum scale system being just a special case of this more general complementarity, as it must be if it is a straightforward consequence of entanglement, which holds, in principle, universally, independently of scale.

Now that we have come to a clear understanding of complementarity and of its being, for Bohr, a direct consequence of entanglement, what further consequences ensue for the way we think about how the seemingly classical macroworld emerges from a quantum microworld. Understanding that will require our taking a close look at another core idea of Bohr's, his doctrine of classical concepts.

Bohr on Classical Concepts

Bohr is well known for having repeatedly argued that, in spite of the fundamental validity of the quantum theory, we must, nevertheless always employ "classical concepts" in the description of experimental setups and experimental outcomes.[4] In one of the better known such remarks, he wrote:

> It is decisive to recognize that, *however far the phenomena transcend the scope of classical physical explanation, the account of all evidence must be expressed in classical terms.* The argument is simply that by the word "experiment" we refer to a situation where we can tell others what we have done and what we have learned and that, therefore, the account of the experimental arrangement and of the results of the observation must be expressed in unambiguous language with suitable application of the terminology of classical physics. (Bohr 1949, 209)

What does Bohr mean by the "terminology of classical physics"? Does this just mean Newtonian mechanics and Maxwellian electrodynamics? There is also the question of the domain of application of classical physical concepts. Does Bohr understand them to apply, as do many, only to macroscopic systems, including measuring apparatus? As I have argued elsewhere, Bohr had something rather more subtle in mind (Howard 1994).

The clearest and most detailed of Bohr's discussions was presented at, of all venues, an anthropology conference in Copenhagen in 1938:

> The elucidation of the paradoxes of atomic physics has disclosed the fact that the unavoidable interaction between the objects and the measuring instruments sets an absolute limit to the possibility of speaking of a behavior of atomic objects which is independent of the means of observation.
>
> We are here faced with an epistemological problem quite new in natural philosophy, where all description of experience has so far been based on the assumption, *already inherent in ordinary conventions of language,* that it is possible to distinguish sharply between the behavior of objects and the means of observation. This assumption is not only fully justified by all everyday experience *but even constitutes the whole basis of classical physics* ... As soon as we are dealing, however, with phenomena like individual atomic processes which,

---

[4] Bohr's doctrine of classical concepts is analyzed in detail in Howard (1994).

due to their very nature, are essentially determined by the interaction between the objects in question and the measuring instruments necessary for the definition of the experimental arrangement, we are, therefore, forced to examine more closely the question of what kind of knowledge can be obtained concerning the objects. In this respect, *we must, on the one hand, realize that the aim of every physical experiment—to gain knowledge under reproducible and communicable conditions—leaves us no choice but to use everyday concepts, perhaps refined by the terminology of classical physics,* not only in all accounts of the construction and manipulation of the measuring instruments but also in the description of the actual experimental results. On the other hand, it is equally important to understand that just this circumstance implies that no result of an experiment concerning a phenomenon which, in principle, lies outside the range of classical physics can be interpreted as giving information about independent properties of the objects. [My italics.] (Bohr 1938, 269)

One wonders what the anthropologists got out of Bohr's lecture! This passage requires careful analysis.

The first sentence says more or less exactly the same as the sentences beginning the introduction of complementarity, namely, that, thanks to entanglement, the observed atomic objects and their means of observation cannot be represented as being mutually independent. Bohr then remarks that this poses a problem, since the contrary assumption that one can distinguish between the behavior of the object and the means of observation is built into "the ordinary conventions of language and constitutes "the whole basis of classical physics." We are in a new land, where the old rules no longer apply, so we have to figure out what kind of knowledge can be obtained about quantum scale objects. Bohr explains that we have no choice but to use everyday, classical concepts—meaning concepts conformable with the assumption of the separability of object and instrument, not their entanglement, so that we can explain unambiguously what we are doing in an experiment and what are the experimental results, meaning, presumably, what values we obtained for the measured properties of the object. But, Bohr concludes, the fact that object and instrument really are entangled means that the measurement cannot be understood as having revealed the "independent properties" of the object.

Appreciate the tension—the dialectical situation, as it were—emphasized here by Bohr: On the one hand, entanglement is a universal fact about interacting systems as described by quantum mechanics, meaning that the instrument and the object form an indissoluble whole (Bohr often uses the term "individuality" in a misleading way to name this holism). On the other hand, objectivity requires the use of classical modes of description in accounting for measurements and their outcomes. How is this dialectical tension to be resolved?

First, a remark. It has been widely assumed that when Bohr wrote about the need to employ classical concepts in the description of experiments and experimental results, he meant their application only to the instrumentation, not to quantum-scale objects, on the assumption that classical physics applies at the macrolevel but not at the microlevel (see, for example, Bokulich 2010). Notice, however, that in the just quoted passage Bohr does not say that classical concepts are simply those of classical mechanics and electrodynamics. He says that what is "classical" is simply the assumption of the mutual independence of object and instrument, their separability.

It is true that this separability principle holds throughout classical physics,[5] but what is inherently "classical" for Bohr is not Newton's three laws of motion or Maxwell's equations, but that underlying assumption of the separability of interacting systems. One question we must, therefore, ask is whether the classical-quantum cut parallels the micro–macro distinction, or whether, perhaps, it is orthogonal to the latter, there being something quantum-like at both the micro- and macrolevel, and something classical at both levels as well.

To understand how Bohr resolves the dialectical tension and to answer that last question of the orientation of the classical-quantum cut, let's look closely at Bohr's own, classic example of a quantum measurement, the double-slit experiment, this from his reply to EPR. Recall the experimental arrangement. A source emits a beam of photons, say, that first pass through a single slit in diaphragm 1 that collimates the beam and then confronts diaphragm 2 with two slits, behind which is a screen coated with a photographic emulsion. In the initial configuration, all components of the apparatus are firmly bolted to the laboratory bench, in which case we observe interference fringes on the screen (Fig. 8.1).

As Bohr explains, in this configuration, that first diaphragm performs, in effect, a measurement of the photon's position along the vertical axis. In a second configuration, one unbolts the first diaphragm from the lab bench, suspending it by a spring, with a pointer and a scale allowing one to track its movement. Curiously, in this configuration, one observes no interference fringes.

In this configuration, the exchange of momentum between the diaphragm and the photon along the vertical axis allows for what is, in effect, a measurement of

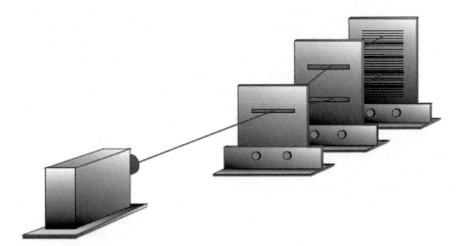

Fig. 8.1 Double-slit experiment, first configuration (Bohr 1949, 219)

---

[5] The assertion of the universal necessity of separability was a fundamental feature of Einstein's philosophy of science and was the real basis of his belief in the incompleteness of quantum mechanics because of its denial of separability through its implying that all previously interacting systems are entangled. See Howard 1985.

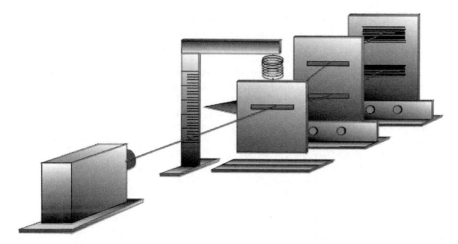

Fig. 8.2 Double-slit experiment, second configuration (Bohr 1949, 220)

the photon's momentum along that axis, which tells us through which slit in the second diaphragm it will pass. As is well known, if one knows through which of those slits the photon passes by placing a detector anywhere along its trajectory, then the interference is destroyed (Fig. 8.2).

What explains the surprising difference between the two configurations? In his reply to EPR, Bohr said the following:

> The principal difference between the two experimental arrangements under consideration is, however, that in the arrangement suited for the control of the momentum of the first diaphragm [movable first diaphragm], this body can no longer be used as a measuring instrument *for the same purpose as in the previous case* [fixed first diaphragm], but must, *as regards its position* relative to the rest of the apparatus, be treated, like the particle traversing the slit, as an object of investigation, in the sense that the quantum mechanical uncertainty relations regarding its position and momentum must be taken explicitly into account. [My italics.] (Bohr 1935, 698)

This is a surprising remark, certainly for those who think that Bohr meant classical concepts to be applied only at the macrolevel and quantum concepts only to be applied at the microlevel, because he here says explicitly that, in the second configuration, with the movable first diaphragm where we measure the momentum of the photon, that both the diaphragm and the photon must be treated quantum mechanically with respect to position. And, by implication, he is saying that, in this configuration, both the photon and the diaphragm are to be treated classically with respect to momentum. In other words, the degrees of freedom of both object and instrument that are not the target of the measurement are to be treated quantum mechanically while the degrees of freedom of both object and instrument that are the target of the measurement must be treated classically. What can all of this possibly mean? In what way can we give a classical account of the photon's momentum? After all, the diaphragm and the photon have, thanks to their interaction, become an entangled pair, and that is a purely quantum mechanical description.

Consider, first, what Bohr says about the coupling of instrument and object degrees of freedom necessary in a measurement (this in a sadly obscure and little known but very important paper):

> We must recognize that a measurement can mean nothing else than the unambiguous comparison of some property of the object under investigation with a corresponding property of another system, serving as a measuring instrument, and for which *this property is directly determinable according to its definition in everyday language or in the terminology of classical physics* .... In the system to which the quantum mechanical formalism is applied, it is of course possible to include any intermediate auxiliary agency employed in the measuring process. Since, however, *all those properties of such agencies which, according to the aim of the measurements have to be compared with the corresponding properties of the object, must be described on classical lines, their quantum mechanical treatment will for this purpose be essentially equivalent with a classical description.* [My italics.] (Bohr 1939, 23–24)

Bohr here reiterates the point that he made in the just quoted remarks about the double-slit experiment in his reply to EPR, namely, that object and instrument degrees of freedom that have to be compared or, shall we say, correlated for the measurement to be the kind of measurement that it is must be described classically. But to that he now adds the perhaps even more puzzling assertion that, since this is true of those degrees of freedom, "their quantum mechanical treatment will for this purpose be essentially equivalent with a classical description." What can that possibly mean? In what sense can a quantum and classical description be "essentially equivalent"? Notice that we are not talking about anything like a correspondence principle, where, in some limit—mass, size, number of degrees of freedom—quantum and classical accounts converge. No we are saying that the quantum mechanical description of the entangled micro–macro pair is essentially equivalent to the classical description of the degrees of freedom that must be correlated in the measurement for it to be the kind of measurement that it is. How can this be?

We need one more clue. We must recall Bohr's often repeated comments about the crucial role of the experimental arrangement in the definition of a quantum phenomenon. In the same paper where Bohr made the cryptic remark about the essential equivalence of the quantum and classical descriptions, he said this about his special way of understanding the term, "phenomenon," in quantum mechanics:

> The essential lesson of the analysis of measurements in quantum theory is thus the emphasis on the necessity, in the account of the phenomena, of taking the whole experimental arrangement into consideration, in complete conformity with the fact that all unambiguous interpretation of the quantum mechanical formalism involves the fixation of the external conditions, defining the initial state of the atomic system concerned and the character of the possible predictions as regards subsequent observable properties of that system. Any measurement in quantum theory can in fact only refer either to a fixation of the initial state or to the test of such predictions, and it is first the combination of measurements of both kinds which constitutes a well-defined phenomenon ... The conditions, which include the account of the properties and manipulation of all measuring instruments essentially concerned, constitute in fact the only basis for the definition of the concepts by which the phenomenon is described. (Bohr 1939, 20)

What work is done by the specification of the experimental arrangement? The specification of the measurement context is the "only basis for the definition of the concepts by which the phenomenon is described." But those concepts are the very "classical" concepts—meaning, now, a description that assumes the separability of object and instrument—that are employed to describe the object and instrument degrees of freedom that must be correlated in a measurement for it to be the kind of measurement that it is. The measurement context defines the kind of classical, separable description that is to be given to those degrees of freedom.

We are ready, now, to begin the assembly of all of these pieces—entanglement, complementarity, classical concepts, and Bohr's somewhat idiosyncratic conception of "phenomena" as involving essentially the specification of the measurement context—into a coherent, unified picture. Here is how Bohr, himself, did it. In the 1949 paper on his discussions with Einstein which was quoted above to first introduce Bohr's doctrine of classical concepts, Bohr said of that doctrine:

> This crucial point ... implies the *impossibility of any sharp separation between the behaviour of atomic objects and the interaction with the measuring instruments which serve to define the conditions under which the phenomena appear.* In fact, the individuality of the typical quantum effects finds its proper expression in the circumstance that any attempt of subdividing the phenomena will demand a change in the experimental arrangement introducing new possibilities of interaction between objects and measuring instruments which in principle cannot be controlled. Consequently, evidence obtained under different experimental conditions cannot be comprehended within a single picture, but must be regarded as *complementary* in the sense that only the totality of the phenomena exhausts the possible information about the objects. (Bohr 1949, 209-210)

Because of entanglement, we cannot, in principle, regard object and instrument as separable, which latter is the essence of a "classical" description, but our need to communicate the results of our experiments unambiguously requires that we do just that, which we can do only if we first restrict that classical description to those object and instrument degrees of freedom that have to be correlated in the measurement for it to be the kind of measurement that it is. We do that restriction through the specification of the experimental context. But different experimental contexts are mutually incompatible, as with that first diaphragm in the double-slit experiment that cannot simultaneously be bolted firmly to the lab bench, when we do a position measurement on the photon, and free to move up and down, when we do a momentum measurement. That physical incompatibility of the measurement contexts entails the complementarity between position and momentum.

For the most part, Bohr always worked out such ideas in words, not equations. Though his brother, Harald, was a prominent mathematician, Bohr, himself, was not all that mathematically gifted and worried that mathematics obscured as much as or more than what it revealed. Nonetheless, one wonders whether Bohr's understanding of entanglement, classical concepts, the centrality of the measurement context, and complementarity can be rendered in more precise mathematical terms, and whether, having done so, we can explain what Bohr meant when he said that, with respect to the degrees of freedom targeted in a specific measurement context, the quantum and classical descriptions will be essentially equivalent?

Context-Dependent Mixtures

Yes, we can do that. Forty years ago, I first proved the relevant theorem (Howard 1979, 382–386), which I termed "a theorem concerning context-dependent mixtures." Consider Bohm's version of the EPR thought experiment (Bohm 1951, 614–623). We imagine a situation in which a diatomic molecule with spin zero decomposes into separate atoms with spin ½ that fly off in opposite directions. L and R, or an electron and a positron resulting from pair production do the same. A Stern-Gerlach apparatus in each wing can be oriented so as to measure the spin of particle L or particle R along any axis orthogonal to the line of flight.

Spin conservation requires perfect anticorrelation, so that, if, for example, particle L has a y-spin of $+\frac{1}{2}$, then particle R must have a y-spin of $-\frac{1}{2}$, and so forth. The pure case density matrix corresponding to the entangled, pure, joint state of L and R before any spin measurement is performed would be:

$$W_{qm} = \frac{1}{2}\left(\left(\left|u^z+(L)\right\rangle\left|u^z-(R)\right\rangle - \left|u^z-(L)\right\rangle\left|u^z+(R)\right\rangle\right)\right.$$
$$\left. + \left(\left\langle u^z+(L)\right|\left\langle u^z-(R)\right| - \left\langle u^z-(L)\right|\left\langle u^z+(R)\right|\right)\right)$$

The theorem then asserts the following. First specify a measurement context in the form of a maximal set of co-measurable joint L and R observables, which, simplified for the case of the Bohm-EPR thought experiment would amount to a specification of the orientation of the Stern-Gerlanch apparatus in each wing. Then, given that specification of a context, there is a constructive procedure whereby one can write down a joint density matrix corresponding not to a pure case but to a proper mixture over joint L and R eigenstates that gives, for observables measurable in that context, exactly the same statistical predictions for correlations between L and R as are given by the pure case density matrix. If we are measuring z-spin in both wings, then the context dependent density matrix is (Fig. 8.3):

$$W_{c1}^{zz} = \left|u^z+(L)\right\rangle\left|u^z-(R)\right\rangle N^z + /N\left\langle u^z+(L)\right|\left\langle u^z-(R)\right|$$
$$+ \left|u^z-(L)\right\rangle\left|u^z+(R)\right\rangle N^z - /N\left\langle u^z-(L)\right|\left\langle u^z+(R)\right|$$

If we are measuring y-spin in both wings, then the context dependent density matrix is:

$$W_{c1}^{yy} = \left|u^y+(L)\right\rangle\left|u^y-(R)\right\rangle N^y + /N\left\langle u^y+(L)\right|\left\langle u^y-(R)\right|$$
$$+ \left|u^y-(L)\right\rangle\left|u^y+(R)\right\rangle N^y - /N\left\langle u^y-(L)\right|\left\langle u^y+(R)\right|$$

And so on for any combination of measurements in L and R. Generally speaking, these context-dependent mixtures do not yield the correct predictions for observables that are not measurable in the stipulated context. I proved the theorem for observables with discrete spectra in 1979. The generalization to observables with continuous

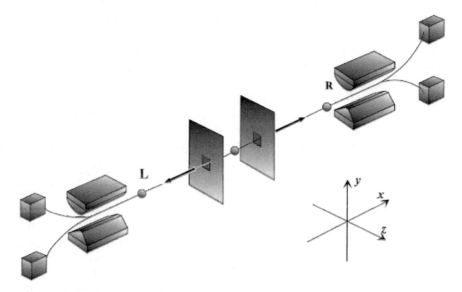

Fig. 8.3 Bohm-EPR thought experiment

spectra was proven by Robert Clifton and Hans Halvorson some twenty years later (Clifton and Halvorson 1999, 2002; Halvorson 2004).

Notice that these context-dependent mixtures are not reduced density matrices representing improper mixtures for, say, particle L, obtained by tracing over particle R's degrees of freedom. These are joint density matrices representing proper mixtures over joint L and R eigenstates. Being such, they can be interpreted as if they represented, *with respect to the degrees of freedom measurable in the stipulated context and only those degrees of freedom*, mutually independent systems. In that respect, they correspond to what Bohr regarded as classical descriptions.

The theorem is general, not restricted to spin. If, instead of looking at spin correlations between coupled atoms, we applied this analysis to the measurement interaction between say, a spin-½ atom and the Stern-Gerlach apparatus plus detectors that measures its spin along the z-axis or the y-axis, then we have a situation strictly analogous to Bohr's account of the double-slit experiment, the atom and its spin replacing the photon and either its position or its momentum along the vertical axis, with the Stern-Gerlach apparatus plus detectors replacing the first diaphragm that is either fixed to the laboratory bench, enabling it to measure the photon's position along the vertical axis, or hanging from a spring, enabling it to measure the photon's momentum along the vertical axis. Those two alternatives for the first diaphragm, fixed or free to move, correspond to the choice of an observable to measure in the left wing of the Bohm-EPR thought experiment, z-spin or y-spin, meaning the orientation of the Stern-Gerlach apparatus plus detectors. Thus, we have a formal explication of Bohr's view that, while the pure state is the proper quantum mechanical account of the post-measurement, system-instrument joint state, we nonetheless can give a classical account of the system and instrument degrees of freedom that have to be correlated for the measurement to be the kind of measurement that it is, with that classical description being "essentially equivalent" to the quantum mechanical description. Finally, as with Bohr's account of the EPR thought experiment, we now also have a fully general account of how complementarity is a direct consequence of entanglement and the doctrine of classical concepts. This is because the measurement contexts that fix the relevant, "classical," context-dependent mixtures over system-instrument eigenstates, are simply, physically incompatible with one another, in the sense that one cannot simultaneously orient the Stern-Gerlach apparatus plus detectors to measure spin along orthogonal axes.

There are several, noteworthy features of this formal explication of Bohr's views on complementarity and classical concepts. First, there is no dualism in the ontology based on scale. This is quantum mechanics all the way up and down. Second, this makes perfect sense of Bohr's contextual notion of a quantum "phenomenon." Third, this way of understanding Bohr, as mentioned, makes perfectly clear the strict equivalence of the quantum and "classical" descriptions with respect to the degrees of freedom involved in a measurement. Fourth, it perfectly reconciles the tension noted by Bohr between the physical fact of entanglement and the requirement that the kind of "classical" description needed for "objectivity," treat object and instrument as if not entangled. A final and noteworthy feature of this way of viewing Bohr's complementarity interpretation of quantum mechanics makes clear why quantum indeterminacy is a consequence of entanglement and contextualism, because, while the context-dependent mixtures preserve strict correlations within a context (hence strict "causality") such strict correlations are not preserved between other degrees of freedom, those not defining the context.

Complementarity and Decoherence Revisited

The challenge in understanding decoherence was to figure out how our applying linear, Schrödinger dynamics to a quantum-scale system interacting with a measuring instrument and millions of environmental degrees of freedom could yield a quantum mechanical pure state for the system-instrument-environment complex that was observationally indistinguishable from a mixture over eigenstates. We now have our answer.

The answer is that applying linear, Schrödinger dynamics to the system-instrument-environment interaction drives the joint, system-instrument-environment state into an entangled, pure, joint state that is observationally indistinguishable from the relevant context-dependent mixture picked out by the measurement context because the pure, joint state and that mixture over joint eigenstates picked out by the measurement context give exactly the same statistical predictions for all observables measurable in that context. We have not merely observational in distinguishability but strict, mathematical equivalence. Thus, environmentally-induced decoherence (which is not really decoherence, but only a semblance thereof) was, all along, the real point toward which Bohr was gesturing with the doctrines of complementarity and classical concepts.

Let us now step back to appreciate some important features of the new view of the quantum–classical relationship afforded by complementarity, properly understood, and decoherence, properly understood, which, we now see, are, for all intents and purposes, one and the same. First, both decoherence and complementarity involve quantum mechanics all the way up and down. Quantum mechanics is the fundamental description of nature at all scales. Second, from both points of view, classicality emerges from fundamental quantum behavior. It is not something fundamentally different in kind. Third, decoherence confirms Bohr's suggestion that classicality, in the guise of seemingly irreversible marks as the outcomes of measurements, arises because of the dissipation of information through the observed system's interaction with the environment and the consequent, practical (not in principle) irreversibility of measurement processes. There is practical irreversibility because the Poincaré cycle times for system-instrument-environment complexes with millions of degrees of freedom are, typically, some multiples of the lifetime of the universe.[6]

The point of view advanced here does not solve one problem that has been raised in the literature on decoherence, the so-called "problem of outcomes" (Schlosshauer 2007, 57–60). This is, simply put, the problem that Schrödinger dynamics applied to the system-instrument-environment complex does not and cannot explain why, when a measurement is performed, a specific, determinate outcome is obtained. The particle is found to be here, not there. The electron's spin along the z-axis is found to be $+\frac{1}{2}$, not $-\frac{1}{2}$. Or so it seems. But three points must be made. First, if we take the physics of decoherence seriously, then, as just explained, the appearance of a seemingly permanent mark on a photographic emulsion or the registration of a click in a Geiger counter is only practically and apparently irreversible. If the system-instrument-environment complex were otherwise held constant indefinitely, and if, *per impossible*, we could wait some multiple of the life of the universe, then the mark would disappear from one place and appear in another and the click would be heard in a different channel. With very tiny systems, those with very short Poincaré cycle times, we see this happening all the time. Indeed, the whole stability of matter depends on it, for that is precisely what is going on with a covalent chemical bond, where the entangled electrons in the valence orbitals are shared by the two atoms, being everywhere and never, even for the briefest moment, in a specific place. Moreover, it has become, of late, a popular physics game to find clever ways of revealing tiny but composite systems in superposition states (see, for example, Kovachy et al. 2015; Proietti et al. 2019).

Second, it is simply not true that, when we measure, we always find the system in a definite eigenstate of the observable of interest. Consider measurements of spin with a Stern-Gerlach apparatus. Thanks to the details of the experimental arrangement, we couple the atom's spin to position degrees of freedom in the apparatus. But those positions are not absolutely sharp points in space. Moreover, for each possible value

---

[6] For a different, but not entirely incompatible perspective on the relationship between complementarity, classical concepts, and decoherence, see Camilleri and Schlosshauer (2015).

Fig. 8.4 Stern-Gerlach
experiment (Gerlach and
Stern 1922, 350)

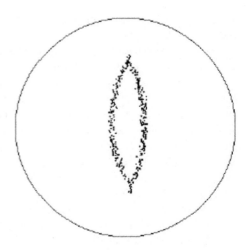

of the atom's spin, there are non-zero probability amplitudes for finding the atom
over a surprising wide range of possible positions. And those amplitudes for spin-
up and spin-down actually overlap. Here is a rendition of the photograph from the
original Stern-Gerlach experiment (Fig. 8.4).

The image on the left is what was seen with the magnetic field turned off, the image
on the right being what was seen with the field turned on. What are we to make of
the regions not in the middle, where the bands are widely separated, but at the ends,
where the two bands merge, at the top and the bottom? Notice that those areas are
just as dark as are the areas where the separation of the bands is greatest. Blow
up the image to resolve the individual dots. What do the dots at the extreme upper
or lower end represent? Spin up? Spin down? You will answer that those extrema
represent atoms that drifted too far this way or that for the magnetic field to have
resolved the spins unambiguously, the field strength attenuating with distance from
the center of the apparatus. But, no. That way of putting it assumes that each atom
traversing the apparatus had a determinate trajectory, which is not possible. Instead,
every atom traversing the apparatus has non-zero amplitudes for every position on
an axis orthogonal both to the line of flight and to the field orientation, and those
amplitudes interfere with one another. One simply cannot say that some specific
atom that seemed to have landed at the upper or lower intersections of the bands
must have been on a trajectory sufficiently off-beam for the field to resolve the spins.
That would not be a well-formulated sentence in the language of the quantum theory.

The third and final point to be made about the failure of the approach to decoherence through context-dependent mixtures to solve the problem of outcomes is, simply,
that decoherence cannot and was not intended to solve the problem of outcomes nor
is it right to expect it do so. Quantum mechanics with the Born rule is a stochastic
theory, so the description afforded by the context-dependent mixtures that give us
the sought-for simulacrum of classicality must be so as well.

Let us return, one last time, to the question posed at the beginning of this discussion. In what precise sense is the pure state into which the decoherence dynamics

drives the system-instrument-environment complex observationally indistinguishable from a mixture over eigenstates. The answer is that that pure state is observationally indistinguishable from the appropriate context-dependent mixture over joint system- instrument-environment eigenstates because, as the theorem proves, that mixture gives exactly the same predictions as the pure state for all observables constituting the context. That is how the illusion of a classical world emerges, ironically, from a thoroughgoing, purely quantum mechanical description of reality.

Acknowledgements This paper has been long in gestation. It began as an invited talk at a workshop on "Classical Concepts and Metaphysical Presuppositions in Quantum Physics," organized by Henrik Zinkernagel in Grenada in 2008. Successive versions of the talk were given at an American Physical Society conference in Raleigh, North Carolina, also in 2008, at a conference on "New Directions in Physics" at the Bucharest Colloquium in Analytic Philosophy in 2013 and as invited lectures at the Niels Bohr Institute in Copenhagen and the Munich Center for Mathematical Philosophy in 2018. I am deeply indebted to the many colleagues who offered helpful commentary on all of those occasions.

A Note on Einstein Citations Correspondence that has already been published in *The Collected Papers of Albert Einstein* (Princeton, NJ: Princeton University Press, 1987-present) is cited by volume number, document number, and page, after the model "CPAE-x, Doc. yyy, z." Unpublished items from Einstein's correspondence are cited by their control index numbers in the Einstein Archive, after the model "EA xx-xxx."

References

Baggott J (2011) The quantum story: a history in 40 moments. Oxford University Press, New York

Beller M, Fine A (1994) Bohr's response to EPR. In: Faye and Folse, pp 11–31

Belousek DW (1996). Einstein's 1927 unpublished hidden-variable theory: its background, context and significance. Stud Hist Philos Mod Phys 27:437–461

Bohm D (1951) Quantum theory. Prentice-Hall, New York

Bohm D (1952a) A suggested interpretation of the quantum theory in terms of hidden variables. I. Phys Rev 85:166–179

Bohm D (1952b) A suggested interpretation of the quantum theory in terms of hidden variables. II Phys Rev 85:180–193

Bohr N (1928) The quantum postulate and the recent development of atomic theory. Nature (suppl.) 121:580–590

Bohr N (1935) Can quantum-mechanical description of physical reality be considered complete? Phys Rev 48:696–702

Bohr N (1938) Natural philosophy and human cultures. In: Comptes Rendus du Congrès International de Science, Anthropologie et Ethnologie. Copenhagen, 1938. Reprinted in Nature, 143 (1939), pp 268–272

Bohr N (1939) The causality problem in atomic physics. In: New theories in physics. International Institute of Intellectual Co-operation, Paris, pp 11–30

Bohr N (1949) Discussion with Einstein on epistemological problems in atomic physics. In: Schilpp PA (ed) Albert Einstein: philosopher-scientist. The Library of Living Philosophers, Evanston, IL, pp 199–241

Bokulich A (2010) Three approaches to the quantum—classical relation: Bohr, Heisenberg, and Dirac. Iyyun: Jerusalem Philos Q 59:3–28

Camilleri K, Schlosshauer M (2015) Niels Bohr as philosopher of experiment: does decoherence theory challenge Bohr's doctrine of classical concepts? Stud Hist Philos Mod Phys 49(2015):73–83

Clifton R, Halvorson H (1999) Maximal Beable Subalgebras of quantum mechanical observables. Int J Theor Phys 38:2441–2484

Clifton R, Halvorson H (2002) Reconsidering Bohr's reply to EPR. In: Placek T, Butterfield J (eds), Non-locality and modality. Kluwer, Dordrecht and Boston, pp 3–18

Crull E (2011) Quantum decoherence and interlevel relations. Ph.D. Dissertation. University of Notre Dame

Einstein A (1905) Über einen die Erzeugung und Verwandlung des Lichtes betreffenden heuristischen Gesichtspunkt. Ann Phys 17:132–148

Einstein A (1925) Quantentheorie des einatomigen idealen Gases. Zweite Abhandlung. Preussische Akademie der Wissenschaften. Physikalisch-mathematische Klasse. Sitzungsberichte, 3–14

Einstein A, Podolsky B, Rosen N (1935) Can quantum-mechanical description of physical reality be considered complete? Phys Rev 47:777–780

Everett H (1957) 'Relartive state' formulation of quantum mechanics. Rev Mod Phys 29:454–462

Faye J, Folse H (eds) (1994) Niels Bohr and contemporary philosophy. Kluwer, Dordrecht

Gerlach W, Stern O (1922) Der experimentelle Nachweis der Richtungsquantelung im Magnetfeld. Z Phys 9:349–352

Ghirardi G, Rimini A, Weber T (1985) A model for a unified quantum description of macroscopic and microscopic systems. In: Accardi L, von Waldenfels W (eds), Quantum probability and applications II: proceedings of a workshop held in Heidelberg, West Germany, Oct 1–5, 1984. Springer, Berlin, pp 223–232

Halvorson H (2004) Complementarity of representations in quantum mechanics. Stud Hist Philos Mod Phys 35:45–56

Heisenberg W (1927) Über den anschaulichen Inhalt der quantentheoretischen Kinematik und Mechanik. Zeitschrift für Physik 172–198

Heisenberg W (1930) The physical principles of the quantum theory. University of Chicago Press, Chicago

Høffding H (1907) Formel logik: til Brug ved Forelaesninger, 5th edn. Gyldendalske Boghandel, Copenhagen

Howard D (1979) Complementarity and ontology: Niels Bohr and the problem of scientific realism in quantum physics. Dissertation. Boston University

Howard D (1990) 'Nicht sein kann was nicht sein darf', or the prehistory of EPR, 1909–1935: Einstein's early worries about the quantum mechanics of composite systems. In: Miller A (ed) Sixty-two years of uncertainty: historical, philosophical, and physical inquiries into the foundations of quantum mechanics. Plenum, New York, pp 61–111

Howard D (1994) What makes a classical concept classical? Toward a reconstruction of Niels Bohr's philosophy of physics. Faye Folse 1994:201–229

Howard D (2004) Who invented the Copenhagen interpretation? A study in mythology. Philos Sci 71:669–682

Howard D (2009) Pauli's 1933 Die allgemeinen Prinzipien der Wellenmechanik. Lecture in the symposium on "A history of quantum physics through the textbooks," History of science society annual meeting, Phoenix, Arizona, 21 Nov 2009

Joos E, Zeh HD (1985) The emergence of classical properties through interaction with the environment. Zeitschrift Für Physik B Condensed Matter 59:223–243

Katsumori M (2011) Niels Bohr's complementarity: its structure, history, and intersections with hermeneutics and deconstruction. Springer, Dordrecht

Károlyházy F (1966) Gravitation and quantum mechanics of macroscopic objects. Il Nuovo Cimento A 42:390–402

Kovachy T et al (2015) Quantum superposition at the half-metre scale. Nature 528:418–530

Kübler O, Zeh HD (1973) Dynamics of quantum correlations. Ann Phys 76:405–418

London F, Bauer E (1939) La Théorie de l'observation en mécanique quantique. Hermann & Cie, Paris

Moore R (1966) Niels Bohr: the man, his science, & the world they changed. Alfred Knopf, New York

Pauli W (1933) Die allgemeinen Principien der Wellenmechanik. In: Geiger H, Scheel K (eds), Handbuch der Physik, 2nd ed. Julius Springer, Berlin, pp 83–272

Penrose R (1986) Gravity and state vector reduction. In: Penrose R, Isham CJ (eds) Quantum concepts of space and time. Oxford University Press, Oxford, pp 129–146

Proietti M et al (2019) Experimental rejection of observer-independence in the quantum world. arXiv:1902.05080

Röseberg U (1992) Niels Bohr. Leben und Werk eines Atomphysikers 1885–1962, 3rd edn. Spektrum, Heidelberg

Schlosshauer M (2007) Decoherence and the quantum-to-classical transition. Springer, Berlin

Schlosshauer M, Camilleri K (2017) Bohr and the problem of the quantum-to-classical transition. In: Faye J, Folse H (eds) Niels Bohr and the philosophy of physics: twenty-first century perspectives. Bloomsbury, London, New York, pp 223–233

Schopenhauer A (1819) Die Welt als Wille und Vorstellung. F. A. Brockhaus, Leipzig

Schrödinger E (1935a) Die gegenwärtige Situation in der Quantenmechanik. Die Naturwissenschaften 23:807–812, 823–828, 844–849

Schrödinger E (1935b) Discussion of probability relations between separated systems. Proc Camb Philos Soc 31:555–662

Schrödinger E (1936) Probability relations between separated systems. Proc Camb Philos Soc 32:446–452

Shimony A (1963) The role of the observer in quantum theory. Am J Phys 31:742–755

von Neumann J (1932) Mathematische Grundlagen der Quantenmechanik. Julius Springer, Berlin

Wallace D (2012) The emergent multiverse: quantum theory according to the Everett interpretation. Oxford University Press, Oxford

Weyl H (1931) Gruppentheorie und Quantenmechanik, 2nd edn. S. Hirzel, Leipzig

Wigner E (1962) Remarks on the mind-body question. In: Good IJ (ed) The scientist speculates. Heinemann, London, pp 284–301

Zeh HD (1970) On the interpretation of measurement in quantum theory. Found Phys 1:69–76

Zeh HD (1973) Toward a quantum theory of observation. Found Phys 3:109–116

Zurek W (1981) Pointer basis of quantum apparatus: into what mixture does the wave packet collapse? Phys Rev D 24:1516–1525

Zurek W (1982) Environment-induced superselection rules. Phys Rev D 26:1862–1880

Chapter 9
Complementarity Between One- and Two-Body Visibilities

Christoph Dittel and Gregor Weihs

Introduction

Complementarity was initially discussed (Bohr 1935, 1949) and first made quantitative at the single particle level (Greenberger and Yasin 1988; Jaeger et al. 1995; Englert 1996). For multiple particles the issue is obviously much more complicated, because there are a myriad possibilities of defining measures of different types of information that might be complementary or not. Yet, the single-particle case can be lifted to the two-particle level with complementarity between one- and two-particle visibility for the scenario where the two particles are subject to separate two-mode interferometers (Jaeger et al. 1993; Schlienz and Mahler 1995). Clearly Mike Horne wanted to expand on this and one of us (GW) distinctly remembers his talk at the conference "Epistemological and Experimental Perspectives on Quantum Physics" (Vienna, Austria, 1998) (Horne 1999) where he tried to extend complementarity to the interference of three particles. The talk showed that there is no such straightforward generalization. In a recent result (Dittel et al. 2019), we were able to derive complementarity and duality relations for multiple, partially distinguishable interfering particles but not in the way envisaged by Mike Horne. Both approaches reveal the crucial role of entanglement, and for tripartite systems it is known that entanglement measures cannot both be faithful and monogamous (Lancien et al. 2016). On the other hand, the situation is rather simple for two two-level systems (qubits) since all measures of entanglement are similar in this case. Under this perspective, we here elucidate a possible connection between the two scenarios at the two-particle level

C. Dittel (✉)
Physikalisches Institut, Albert-Ludwigs-Universität Freiburg, Hermann-Herder-Str. 3, 79104 Freiburg, Germany
e-mail: christoph.dittel@physik.uni-freiburg.de

G. Weihs
Institut für Experimentalphysik, Universität Innsbruck, Technikerstr. 25, 6020 Innsbruck, Austria
e-mail: gregor.weihs@uibk.ac.at

© Springer Nature Switzerland AG 2021
G. Jaeger et al. (eds.), *Quantum Arrangements*, Fundamental Theories of Physics 203, https://doi.org/10.1007/978-3-030-77367-0_9

(a)   (b)   (c)

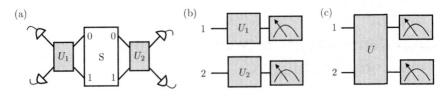

Fig. 9.1 Schematic experimental settings to probe the complementarity of one- and two-body visibilities. **a** A source S emits two particles, which are possibly entangled in their mode occupation, with modes labelled 0 and 1. The particles leave the source in opposite directions, are separately transformed according to U_1 and U_2, and measured in the output modes, respectively. **b** In a similar setting as in (**a**), two possibly entangled qubits, labelled 1 and 2, undergo local unitary transformations with a subsequent measurement in the computational basis. **c** In a similar setting as in (**b**), two qubits are transformed according to a common (global) two-qubit unitary transformation U

with special emphasis on the role of entanglement. The original scheme involves a two-particle source, with the particles sent in opposite directions, and possibly entangled in their occupation in two distinct modes [see Fig. 9.1a]. The particles are then separately transformed according to two-mode unitary transformations [U_1 and U_2 in Fig. 9.1a], and subsequently measured in the output modes. Thereby, one- and two-body visibilities are obtained after an optimization procedure over the local unitary transformations U_1 and U_2, which were shown to be in a complementary relation (Jaeger et al. 1993, 1995).

By considering the mode occupation of each particle as a two-level system, this experimental scheme is similar to two possibly entangled qubits undergoing local unitary transformations followed by a measurement in the computational basis [compare Fig. 9.1a, b]. Now, by allowing for common—i.e. global—two-qubit unitaries instead [see Fig. 9.1c], we can ask whether similar complementarity relations between one- and two-body visibilities can be obtained.

One-Body Visibility

To set the stage, let us consider a pure state $|\Psi\rangle$ of two qubits, which we write in its Schmidt decomposition, $|\Psi\rangle = \sum_{j=0}^{1} \sqrt{\lambda_j} |\eta_j\rangle |\xi_j\rangle$, with $\lambda_j \in [0, 1]$, $\lambda_0 + \lambda_1 = 1$, and $\{|\eta_j\rangle\}_{j=0}^{1}$ and $\{|\xi_j\rangle\}_{j=0}^{1}$ an orthonormal basis of the first and second qubit, respectively. Note that w.l.o.g. we assume $\lambda_0 \geq \lambda_1$, such that $1/2 \leq \lambda_0 \leq 1$. The reduced state of the first (resp. second) qubit is then obtained from the two-qubit density operator $\rho = |\Psi\rangle\langle\Psi|$ by tracing over the Hilbert space of the second (resp. first) qubit, $\rho_1 = \sum_{j=0}^{1} \lambda_j |\eta_j\rangle\langle\eta_j|$ (resp. $\rho_2 = \sum_{j=0}^{1} \lambda_j |\xi_j\rangle\langle\xi_j|$), where λ_0 and λ_1 appear as the eigenvalues of ρ_1 (resp. ρ_2).

First, let us consider the reduced state ρ_1 of the first qubit under local unitary transformations U_1 as shown in Fig. 9.1b. In this setting, the one-body visibility is commonly written as

$$v_1 = \frac{p_1^{\max}(0) - p_1^{\min}(0)}{p_1^{\max}(0) + p_1^{\min}(0)}, \tag{9.1}$$

with $p_1^{\max}(0) = \max_{U_1} \mathrm{Tr}(|0\rangle\langle 0|U_1\rho_1 U_1^\dagger)$ (resp. $p_1^{\min}(0) = \min_{U_1} \mathrm{Tr}(|0\rangle\langle 0|U_1\rho_1 U_1^\dagger)$) the maximal (resp. minimal) probability to find the first qubit in $|0\rangle$ after an optimization over U_1. Now, for $U_1^{\max}$ the unitary leading to $p_1^{\max}(0)$, we see that the unitary $\sigma_x U_1^{\max}$, with the Pauli matrix σ_x flipping $|0\rangle$ and $|1\rangle$, gives rise to $p_1^{\max}(1) = p_1^{\max}(0)$. Moreover, with $p_1^{\min}(0) = 1 - p_1^{\max}(1)$, we have $p_1^{\min}(0) = 1 - p_1^{\max}(0)$, such that Eq. (9.1) becomes

$$v_1 = 2p_1^{\max}(0) - 1. \tag{9.2}$$

Under this perspective, we may also express the one-body visibility (9.1) in terms of the *Kolmogorov distance* (or L_1 *distance*) (Nielsen and Chuang 2010)

$$D(P_1, P_1^{\mathrm{mix}}) = \frac{1}{2}\sum_{j=0}^{1}\left|p_1(j) - \frac{1}{2}\right| \tag{9.3}$$

between the probability distributions $P_1 = \{p_1(0), p_1(1)\}$ and $P_1^{\mathrm{mix}} = \{1/2, 1/2\}$. The latter is obtained from a maximally mixed one-qubit state $\rho_1^{\mathrm{mix}} = 1/2\sum_{j=0}^{1}|\eta_j\rangle\langle\eta_j|$. In particular, with the help of Eqs. (9.2) and (9.3), we find

$$v_1 = \max_{U_1} 2\,D(P_1, P_1^{\mathrm{mix}}) \tag{9.4}$$

$$= \max_{U_1}\sum_{j=0}^{1}\left|\mathrm{Tr}\left(|j\rangle\langle j|U_1(\rho_1 - \rho_1^{\mathrm{mix}})U_1^\dagger\right)\right|. \tag{9.5}$$

Here, the maximum is reached if $U_1^\dagger|0\rangle\langle 0|U_1$ and $U_1^\dagger|1\rangle\langle 1|U_1$ project onto the eigenstates of $\rho_1 - \rho_1^{\mathrm{mix}}$, e.g. $U_1^\dagger|0\rangle = |\eta_0\rangle$ and $U_1^\dagger|1\rangle = |\eta_1\rangle$, such that $p_1^{\max}(0) = \lambda_0$. Together with Eq. (9.2), we then arrive at

$$v_1 = 2\lambda_0 - 1. \tag{9.6}$$

This expression allows us to write v_1 as the normalized purity (Dittel et al. 2019) of the reduced one-qubit state,

$$v_1 = \sqrt{2\mathrm{Tr}\left(\rho_1^2\right) - 1}, \tag{9.7}$$

or in terms of the concurrence $\mathcal{C} = 2\sqrt{\lambda_0\lambda_1}$ Wootters (1998) of ρ, which quantifies the entanglement between the qubits,

$$v_1 = \sqrt{1 - \mathcal{C}^2}. \tag{9.8}$$

Accordingly, maximally entangled qubits, for which $C = 1$, imply a vanishing visibility $v_1 = 0$, and separable qubits, for which $C = 0$, lead to $v_1 = 1$. Given that the concurrence C is an entanglement measure, it is apparent from Eq. (9.8) that the one-body visibility v_1 quantifies the separability of the qubits, and, thus, stands in a complementary relation to the entanglement between the qubits. In this regard, we now focus on visibility measures of two-body correlators that quantify the amount of entanglement under an optimization of global two-qubit unitary transformations U as illustrated in Fig. 9.1c.

Two-Body Visibilities

Previous works (Jaeger et al. 1993, 1995; Schlienz and Mahler 1995; Horne 1999) considered the experimental schemata of Fig. 9.1a or b, with the two-body visibility

$$v_{12} = \frac{\bar{p}^{\text{loc.max}}(0,0) - \bar{p}^{\text{loc.min}}(0,0)}{\bar{p}^{\text{loc.max}}(0,0) + \bar{p}^{\text{loc.min}}(0,0)} \tag{9.9}$$

obtained under optimization of the two-body correlator

$$\bar{p}(j,k) = p(j,k) - p_1(j)p_2(k) + 1/4 \tag{9.10}$$

with respect to local unitary transformations U_1 and U_2 [indicated by the superscript loc.max and loc.min]. In this case, $p(j,k) = \text{Tr}(|j,k\rangle\langle j,k|(U_1 \otimes U_2)\rho(U_1 \otimes U_2)^{\dagger})$ is the probability to measure the two-qubit state $|j,k\rangle \equiv |j\rangle |k\rangle$, and the constant $1/4$ is needed in order that $\bar{p}(j,k) \geq 0$ and $\sum_{j,k=0}^{1} \bar{p}(j,k) = 1$, such that $\bar{P} = \{\bar{p}(j,k)\}_{j,k=0}^{1}$ has the properties of a probability distribution. As shown in (Jaeger et al. 1995), Eq. (9.9) results in $v_{12} = C$, such that, by Eq. (9.8), we have

$$v_1^2 + v_{12}^2 = 1. \tag{9.11}$$

Let us note that in (Jaeger et al. 1993) the unitaries U_1 and U_2 are restricted to transformations corresponding to a balanced beam splitter, having equal transmittivity and reflectivity, together with a phase shifter in one of the input modes. Under these restrictions, Eq. (9.11) becomes (Jaeger et al. 1993, 1995)

$$v_1^2 + v_{12}^2 \leq 1. \tag{9.12}$$

By performing the maximization in Eq. (9.9) over *global* two-body unitary transformations U instead of local unitaries of the form $U_1 \otimes U_2$ [see Fig. 9.1c], one can show that $v_{12} = 1$, independently of the initial two-body state ρ. That is, v_{12} does not provide any information about ρ once we allow for global two-body unitaries U.

Nonetheless, by modifying the correlators (9.10) as

$$\bar{c}(j, k) = p(j, k) - p^{\text{sep}}(j, k) + 1/4, \tag{9.13}$$

we can construct entanglement-sensitive two-body visibility measures of the form (9.9), with $\bar{c}(j, k)$ coinciding with $\bar{p}(j, k)$ from Eq. (9.10) in the case of local unitary transformations. While here, the usual probability $p(j, k) = \text{Tr}(|j, k\rangle\langle j, k| U\rho U^{\dagger})$ is obtained from the two-body state ρ, which possibly involves entanglement between the qubits, $p^{\text{sep}}(j, k) = \text{Tr}(|j, k\rangle\langle j, k| U\rho^{\text{sep}}U^{\dagger})$ results from the separable, and, thus, unentangled state $\rho^{\text{sep}} = \rho_1 \otimes \rho_2$. From this perspective, the correlators $\bar{c}(j, k)$ are sensitive to the state's separability, and, as will be shown further down, can be utilized to measure the entanglement between the qubits. Note that again the constant $1/4$ ensures $\bar{c}(j, k) \geq 0$, and the set $\bar{C} = \{\bar{c}(j, k)\}_{j,k=0}^{1}$ has the properties of a probability distribution. Moreover, let us recall that ρ_1 (resp. ρ_2) is obtained from ρ by tracing out the second (resp. first) qubit. In the laboratory, the state $\rho^{\text{sep}} = \rho_1 \otimes \rho_2$ can then be prepared, for example, by assembling the two-qubit system from two separate copies of ρ, with one qubit taken from each copy.

Let us first start out from a similar definition for the visibility as in Eq. (9.9), and consider

$$\tilde{w}_{12} = \frac{\bar{c}^{\text{max}}(0, 0) - \bar{c}^{\text{min}}(0, 0)}{\bar{c}^{\text{max}}(0, 0) + \bar{c}^{\text{min}}(0, 0)}, \tag{9.14}$$

with the optimization performed over global two-qubit unitaries U [cf. Fig. 9.1c]. In order to obtain an expression for $\tilde{w}_{12}$ in terms of the eigenvalue λ_0 [i.e. similar to Eq. (9.6)], we consider

$$\bar{c}^{\text{max}}(0, 0) = \max_{U}[p(0, 0) - p^{\text{sep}}(0, 0) + 1/4]$$
$$= \max_{U}\left[\text{Tr}\left(|00\rangle\langle 00| U(\rho - \rho_1 \otimes \rho_2)U^{\dagger}\right)\right] + 1/4. \tag{9.15}$$

Using the spectral decomposition $\rho - \rho_1 \otimes \rho_2 = \sum_{j=0}^{3} \alpha_j |\alpha_j\rangle\langle\alpha_j|$, with not necessarily positive eigenvalues α_j and their corresponding eigenvectors $|\alpha_j\rangle$, we see that the maximum in (9.15) is obtained if $U^{\dagger}|00\rangle\langle 00|U$ projects onto the subspace spanned by the eigenvector(s) corresponding to the maximal eigenvalue(s) α^{max}, i.e. for $U^{\dagger}|00\rangle = |\alpha^{\text{max}}\rangle$ we have $\bar{c}^{\text{max}}(0, 0) = \alpha^{\text{max}} + 1/4$. Similar, for $\bar{c}^{\text{min}}(0, 0)$ in Eq. (9.14), the minimization gives rise to the minimal eigenvalue(s) α^{min}, leading to $\bar{c}^{\text{min}}(0, 0) = \alpha^{\text{min}} + 1/4$. Therefore, we simply have to evaluate the eigenvalues of $\rho - \rho_1 \otimes \rho_2$, for which we find

$$\alpha_0 = \alpha_1 = -\lambda_0\lambda_1, \quad \alpha_2 = \lambda_0\lambda_1 + \sqrt{\lambda_0\lambda_1}, \quad \alpha_3 = \lambda_0\lambda_1 - \sqrt{\lambda_0\lambda_1}. \tag{9.16}$$

Under consideration of $\lambda_0\lambda_1 \leq 1/4$, we have $\alpha^{\text{max}} = \alpha_2$ and $\alpha^{\text{min}} = \alpha_3$, such that Eq. (9.14) becomes

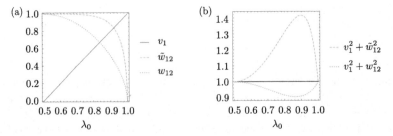

Fig. 9.2 Complementary behaviour of the visibility measures. **a** The one-body visibility v_1 (full blue line) monotonously increases, and the two-body visibilities $\tilde{w}_{12}$ (dashed orange line) and w_{12} (dotted green line) monotonously decrease as a function of the eigenvalue λ_0. This illustrates the complementary behaviour between v_1 and $\tilde{w}_{12}$ (resp. w_{12}). **b** The sum of squared visibility measures $v_1^2 + \tilde{w}_{12}^2$ (resp. $v_1^2 + w_{12}^2$) is shown by a dashed orange (resp. dotted green) line as a function of λ_0, and, in accordance with inequality (9.19) (resp. (9.26)), bounded from below (resp. above) by unity (full black line)

$$\tilde{w}_{12} = \frac{2\sqrt{\lambda_0 \lambda_1}}{2\lambda_0 \lambda_1 + 1/2}. \tag{9.17}$$

This can also be expressed in terms of the concurrence $\mathcal{C}$ of the two-body state ρ, reading

$$\tilde{w}_{12} = \frac{2\mathcal{C}}{\mathcal{C}^2 + 1}. \tag{9.18}$$

From Eqs. (9.17) and (9.18) we see that $0 \leq \tilde{w}_{12} \leq 1$ measures the entanglement between the qubits, with its lower bound reached for separable qubits, i.e. $\tilde{w}_{12} = 0$ for $\mathcal{C} = 0$, and its upper bound reached for maximally entangled qubits, i.e. $\tilde{w}_{12} = 1$ for $\mathcal{C} = 1$. Therefore, $\tilde{w}_{12}$ behaves complementary to the single-body visibility v_1 from Eq. (9.6) as highlighted by their opposite monotonicity as a function of λ_0 shown in Fig. 9.2a. At the same time, however, the visibilities v_1 and $\tilde{w}_{12}$ do *not* satisfy the complementary relation of the form (9.12), instead,

$$v_1^2 + \tilde{w}_{12}^2 \geq 1. \tag{9.19}$$

This inequality results from plugging $\tilde{w}_{12} \geq \mathcal{C}$ [see Eq. (9.18)] into Eq. (9.8), and is graphically illustrated in Fig. 9.2b. While Eq. (9.19) at first seems counterintuitive for two normalized measures that behave in a complementary way, it highlights that satisfying the relation of the form (9.12) is not a sufficient and not even a necessary condition for two measures to behave in a complementary way to each other. Nevertheless, in the following we provide an alternative visibility measure, which we will show to satisfy the form (9.12).

The definitions of the visibilities v_{12} and $\tilde{w}_{12}$ from Eqs. (9.9) and (9.14) are motivated by their similarity to the usual one-body visibility v_1 from Eq. (9.1).

However, as shown in Eq. (9.4), v_1 can likewise be considered as the Kolmogorov distance to the outcome obtained from a maximally mixed state. In view of the two-body correlators (9.13) after global unitary transformations, Eq. (9.4) then motivates the definition of the visibility

$$w_{12} = \frac{4}{3} \max_U D(\bar{C}, \bar{C}^{\text{sep}}), \tag{9.20}$$

with $\bar{C}^{\text{sep}} = \{1/4, 1/4, 1/4, 1/4\}$ the correlator distribution in the case of a separable state $\rho = \rho^{\text{sep}}$. Note that the factor 4/3 arises in order for w_{12} to be normalized. This will get apparent further below. In consideration of the correlators (9.13), w_{12} from Eq. (9.20) is equivalent to

$$w_{12} = \frac{4}{3} \max_U D(P, P^{\text{sep}}), \tag{9.21}$$

with $P = \{p(j,k)\}_{j,k=0}^1$ and $P^{\text{sep}} = \{p^{\text{sep}}(j,k)\}_{j,k=0}^1$ the probability distribution obtained from ρ and ρ^{sep}, respectively. Let us now inspect Eq. (9.21). By plugging in the Kolmogorov distance $D(P, P^{\text{sep}}) = 1/2 \sum_{j,k=0}^1 |p(j,k) - p^{\text{sep}}(j,k)|$, we obtain

$$w_{12} = \frac{2}{3} \max_U \sum_{j,k=0}^1 \left| \text{Tr} \left(|j,k\rangle\langle j,k| U(\rho - \rho_1 \otimes \rho_2) U^\dagger \right) \right|. \tag{9.22}$$

Here, the maximum is reached for $U^\dagger$ rotating the computational basis $\{|j,k\rangle\}_{j,k=0}^1$ to the eigenbasis $\{|\alpha_j\rangle\}_{j=0}^3$ of $\rho - \rho_1 \otimes \rho_2$, e.g. if $U^\dagger |00\rangle = |\alpha_0\rangle$, $U^\dagger |01\rangle = |\alpha_1\rangle$, $U^\dagger |10\rangle = |\alpha_2\rangle$, and $U^\dagger |11\rangle = |\alpha_3\rangle$. Equation (9.22) then becomes $w_{12} = 2/3 \sum_{j=0}^3 |\alpha_j|$, and, with the eigenvalues (9.16), we arrive at

$$w_{12} = \frac{4}{3} \left(\lambda_0 \lambda_1 + \sqrt{\lambda_0 \lambda_1} \right). \tag{9.23}$$

Let us note that we can further express w_{12} in terms of the fidelity $F(\rho_1, \rho_1^{\text{mix}}) = \sum_{j=0}^1 \sqrt{\lambda_j/2}$ of ρ_1 and the maximally mixed one-qubit state ρ_1^{mix} [see below Eq. (9.3)],

$$w_{12} = \frac{4}{3} \left[F^4(\rho_1, \rho_1^{\text{mix}}) - \frac{1}{4} \right], \tag{9.24}$$

where $1/4 \leq F^4(\rho_1, \rho_1^{\mathrm{mix}}) \leq 1$, as well as in terms of the concurrence $\mathcal{C}$ of ρ,

$$w_{12} = \frac{1}{3}\left(\mathcal{C}^2 + 2\mathcal{C}\right). \tag{9.25}$$

Equations (9.23)–(9.25) show that the two-body visibility w_{12} is a normalised measure of the entanglement between the qubits, with $w_{12} = 0$ for separable two-qubit states and $w_{12} = 1$ for maximally entangled states. As shown in Fig. 9.2a, w_{12} monotonously decreases for increasing λ_0, obeying a complementary behaviour to the single-body visibility v_1 from Eq. (9.1). Furthermore, by plugging $w_{12} \leq \mathcal{C}$ [see Eq. (9.25)] into Eq. (9.8), we find these measures to satisfy

$$v_1^2 + w_{12}^2 \leq 1, \tag{9.26}$$

which is graphically illustrated in Fig. 9.2b. There one can also see that the inequality in (9.26) saturates only if v_1 or w_{12} equals unity. In summary, the one- and two-body visibilities v_1 and w_{12} satisfy the usual relation (9.26) associated with complementary measures, which, in the present case, can simply be interpreted in terms of entanglement: the one-body visibility v_1 measures the separability, and the two-body visibility w_{12} the entanglement between both qubits. The more separable the qubits, the less entangled they are, and vice versa.

Discussion and Conclusion

In the literature, quantitative expressions for the complementarity between two measures are often provided in terms of inequalities of the form (9.12). One type of these celebrated inequalities connects one- and two-body visibilities obtained after local unitary transformations of two entangled two-level systems (Jaeger et al. 1993, 1995; Schlienz and Mahler 1995; Horne 1999). Here we went one step further and considered *global* two-body transformations. We introduced two different two-body visibilities $\tilde{w}_{12}$ and w_{12}, and showed that both behave complementary to the usual one-body visibility v_1 in the sense of an opposite monotonicity behaviour. This can ultimately be understood in terms of entanglement: v_1 measures the separability, and both $\tilde{w}_{12}$ and w_{12} the amount of entanglement of the two-body system under consideration. However, we found that only w_{12} satisfies the celebrated complementarity relation $v_1^2 + w_{12}^2 \leq 1$. For $\tilde{w}_{12}$ on the other hand, we even showed that $v_1^2 + \tilde{w}_{12}^2 \geq 1$, although both visibility measures v_1 and $\tilde{w}_{12}$ are normalised and complementary to each other. While this inequality appears unexpected at first sight, it shows that satisfying a complementarity relation of the form (9.12) is not a necessary condition, and, to keep in mind, does not suffice to speak of a complementary behaviour between two measures. It is worth mentioning that it is feasible to measure the here obtained interrelations (9.19) and (9.26) between one- and two-body visibilities experimentally with state-of-the-art technology on diverse experimental platforms, ranging from trapped ion systems to superconducting quantum circuits.

To finish, let us comment on Mike Horne's vision of a three-body complementarity relation, possibly in the form $v_1^2 + v_{12}^2 + v_{123}^2 \leq 1$, with v_{123} a three-body visibility. This relation was disproven by himself (Horne 1999) for an extension of the correlators (9.10) to three parties together with a three-body visibility measure similar to Eq. (9.9). From our above discussion, however, it is clear that such a three-body complementarity relation is preferably addressed in terms of entanglement between the constituents. A promising approach may involve visibility measures in terms of distances between correlator distributions rather than visibilities of single correlators, and possibly two- and three-body correlators similar to those from Eq. (9.13). Yet, a straightforward extension of our results is not possible since we started out from the Schmidt decomposition of a two-body state, which, in general, cannot be generalised to more than two parties. Nonetheless, we are confident that the here established framework provides a promising route to a quantitative study of complementarity in multi-partite systems.

Acknowledgements C.D. would like to thank Giulio Amato and Eric Brunner for fruitful discussions. G.W. would like to thank Barbara Kraus for making us aware of the relation to entanglement monogamy. This work was supported by the Austrian Science Fund (FWF), project nos. I2562 and F7114.

References

Bohr N (1949) The library of living philosophers, Volume 7. Albert Einstein: Philosopher-Scientist. In: Schilpp (ed) Open Court, pp 199–241

Bohr N (1935) Can quantum-mechanical description of physical reality be considered complete? Am Phys Rev 48(8):696–702. https://doi.org/10.1103/PhysRev.48.696

Dittel C, Dufour G, Weihs G, Buchleitner A (2019) arXiv:1901.02810v2

Englert B-G (1996) Fringe visibility and which-way information: an inequality. Phys Rev Lett 77(11):2154–2157. https://doi.org/10.1103/PhysRevLett.77.2154

Greenberger DM, Yasin A (1988) Simultaneous wave and particle knowledge in a neutron interferometer. Phys Lett A 128(8):391–394. https://doi.org/10.1016/0375-9601(88)90114-4

Horne M (1999) Epistemological and experimental perspectives on quantum physics. In: Greenberger D, Reiter WL, Zeilinger A (eds) Springer Netherlands, Dordrecht, pp 211–220

Jaeger G, Horne MA, Shimony A (1993) Complementarity of one-particle and two-particle interference. Phys Rev A 48(2):1023–1027. https://doi.org/10.1103/PhysRevA.48.1023

Jaeger G, Shimony A, Vaidman L (1995) Two interferometric complementarities. Phys Rev A 51(1):54–67. https://doi.org/10.1103/PhysRevA.51.54

Lancien C, Di Martino S, Huber M, Piani M, Adesso G, Winter A (2016) Should Entanglement Measures be Monogamous or Faithful? Phys Rev Lett 117(6):060501. https://doi.org/10.1103/PhysRevLett.117.060501

Nielsen MA, Chuang IL (2010) Quantum computation and quantum information: 10th anniversary edition, 10th edn. Cambridge University Press, New York, NY, USA

Schlienz J, Mahler G (1995) Description of entanglement. Phys Rev A 52(6):4396–4404. https://doi.org/10.1103/PhysRevA.52.4396

Wootters WK (1998) Entanglement of formation of an arbitrary state of two qubits. Phys Rev Lett 80(10):2245–2248. https://doi.org/10.1103/PhysRevLett.80.2245

Chapter 10
GHZ, As Seen Through the Looking Glass

Mordecai Waegell and P. K. Aravind

The GHZ paper (Greenberger et al. 1990) was a seismic event in the field of quantum foundations whose reverberations still echo down to the present. In this article, contributed in Mike Horne's memory, we show how a proof of the Kochen-Specker (KS) theorem (Specker 1987) by Mermin (1993) based on the three-qubit GHZ observables can be transmuted into a proof of the same theorem by Kernaghan and Peres (1995) based on the eigenstates of those observables. We then review some of the later developments that followed from this connection.

Although GHZ (Greenberger et al. 1990) originally gave their inequality-free proof of Bell's theorem (Bell 1964) using a system of four qubits, Mermin (1990) later recast it in terms of three qubits and also showed how the associated observables could be used to give a proof of the KS theorem. Mermin's proof (Mermin 1993) is based on the observables in Fig. 10.1 and runs as follows. The four observables along each edge of the pentagram form a mutually commuting set, with their product being III along the four light edges and $-III$ along the dark edge. A noncontextual hidden variables theory must assign a $+1$ or a -1 to each of the observables in such a way that the product of the values along the light edges is $+1$ while that along the dark edge is -1. However this is impossible because the product of the products must at the same time be both -1 (because four of the products are $+1$ while one is -1) and $+1$ (because each of the values occurs twice over the products), and this impossibility proves the KS theorem. The brevity of this proof is remarkable, considering the difficulty of many of the proofs that preceded it.

M. Waegell
Institute for Quantum Studies, Chapman University, Orange, CA, USA
e-mail: waegell@chapman.edu

P. K. Aravind (✉)
Physics Department, Worcester Polytechnic Institute, Worcester, MA 01609, USA
e-mail: paravind@wpi.edu

© Springer Nature Switzerland AG 2021
G. Jaeger et al. (eds.), *Quantum Arrangements*, Fundamental Theories of Physics 203,
https://doi.org/10.1007/978-3-030-77367-0_10

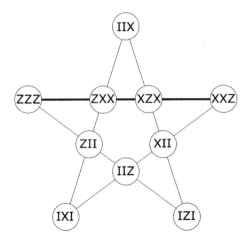

Fig. 10.1 The ten GHZ-Mermin three-qubit observables, arranged along the edges of a pentagram. The product of the observables along each of the light edges is III and along the dark edge is $-III$, where X, Y, Z and I are the Pauli and identity operators of the qubits

Mermin's feat was achieved by shifting the gaze from vectors in Hilbert space to the observables of a small number of qubits. Kernaghan and Peres, in their variation on Mermin's proof, shifted the gaze back from observables to states, seemingly complicating the problem once again, but they did so with a delightful twist that brought back some simplicity with it. Admittedly their proof is not as simple as Mermin's, but bears an interesting relation to it. The best way we can describe the relation is to say that it is a sort of "Looking Glass" version of Mermin's proof, strangely distorted to be sure, but nevertheless quite fascinating and appealing in its own right. Let us try to convince you of this.

To follow Kernaghan and Peres into the Looking Glass world, we must shift our attention from the sets of commuting observables in Fig. 10.1 to their simultaneous eigenstates, shown in Table 10.1. The eight states in any row of the table form a mutually orthogonal and complete set that we will refer to as a "pure" basis (the reason for this adjective will become clear shortly). If one looks closely at the pure bases in Table 10.1, one sees that any two of them have their states split into the same two four-dimensional subspaces by the observable they have in common; for example, the states 1, 2, 3, 4 of the first basis, which have eigenvalue +1 for the observable ZII, occupy the same subspace as the states 9, 10, 11, 12 of the second basis which also have the same eigenvalue for this observable, while the states 5, 6, 7, 8 and 13, 14, 15, 16 occupy the orthogonal subspace. One can therefore "mix and match" the states of these pure bases to form two new bases, which we will term "hybrid" bases, to distinguish them from the pure bases that give rise to them. The five pure bases in Table 10.1 give rise to twenty hybrids, with each hybrid inheriting half its states from each of the pure bases that mated to produce it. We will refer to

Table 10.1 The 40 Kernaghan-Peres states, numbered 1 to 40, shown as simultaneous eigenstates of the sets of commuting observables in Fig. 10.1. Each row shows the eigenstates of one of the commuting sets, with the states being ordered so that their eigenvalue signatures are as indicated at the tops of the columns. A negative sign has been attached to the last observable in the last row to make the eigenvalue signatures come out right

Observables	++++	++ −−	+ − + −	+ − − +	− + + −	− + − +	− − + +	− − − −
$ZII, IZI,$ IIZ, ZZZ	1	2	3	4	5	6	7	8
$ZII, IXI,$ IIX, ZXX	9	10	11	12	13	14	15	16
$XII, IZI,$ IIX, XZX	17	18	19	20	21	22	23	24
$XII, IXI,$ IIZ, XXZ	25	26	27	28	29	30	31	32
$ZZZ, ZXX,$ $XZX,$ $-XXZ$	33	34	35	36	37	38	39	40

the two hybrids resulting from a mating as the "complements" of each other, and to the two together as a complementary pair.

The 25 bases that arise from the Kernaghan-Peres states are shown in Fig. 10.2, with the five pure bases at the vertices of a pentagon and the hybrids at the ends of extensions of its sides and diagonals. This, then, is the Looking Glass equivalent of Fig. 10.1. But how does it help to prove the KS theorem?

Proving the KS theorem requires showing that it is impossible to assign a 0 or 1 to each of the Kernaghan-Peres states in such a way that each of the 25 bases in Fig. 10.2 has exactly one state assigned the value 1 in it. This might seem like a tedious task, and at this point you might be tempted to say "enough" and walk away, but hang on just a bit longer and we'll show you that the answer is right before us, simply waiting to be uncovered.

Let us attach the label U, V, W, X or Y to the pure basis consisting of the states 1–8, 9–16, 17–24, 25–32 or 33–40, respectively. Each pure basis gives rise to four complementary pairs of hybrids, namely, the ones lying at the ends of the four lines passing through it. Let us term the set of four hybrids obtained by picking one member from each of these complementary pairs a **quartet**. Any pure basis has $2^4 = 16$ quartets associated with it.

Let us call a quartet even or odd (and label it by the letter E or O) if every state of the associated pure basis occurs an even (or odd) number of times in it. For example, the quartet of U consisting of the hybrids (1 2 3 4 13 14 15 16), (1 2 5 6 19 20 23 24), (1 3 5 7 26 28 30 32) and (1 4 6 7 37 38 39 40) is even because state 1 occurs four times in it, states 2, 3, 4, 5, 6 and 7 twice each and state 8 not at all. It is not hard to deduce from the eigenvalue signatures in Table 10.1 that every quartet must be either even or odd, but here is an easy way of seeing it.

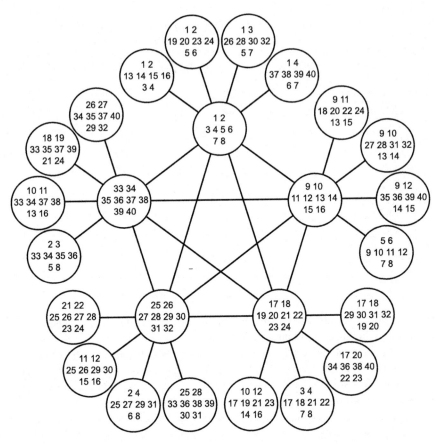

Fig. 10.2 The 25 bases (5 pure and 20 hybrid) formed by the Kernaghan-Peres states. The pure bases are the ones in the inner ring, while the line running through any pair of pure bases ends in the hybrids produced by them

Consider the quartet of U mentioned in the previous paragraph. All the other quartets of U can be obtained from this one by complementing one or more of its hybrids. However complementing a hybrid causes four of the states of U to appear once less in the new quartet and the other four states to appear once more, thus changing the parity of the quartet from E to O or vice-versa. From this it follows that exactly half of the quartets of U are even and the other half odd. The same is true of the quartets associated with the other pure bases as well.

Let us use the term **decuplet** for the set of ten hybrids obtained by picking one member from each complementary pair of hybrids. There are $2^{10} = 1024$ decuplets altogether. Another way of generating a decuplet is to pick one quartet associated with each of the pure bases in such a way that no quartet has a basis that is the complement of a basis in any of the other quartets. If we build up a decuplet in this way, the first quartet can be picked in 2^4 ways, the second in 2^3 ways, the third in 2^2 ways, the fourth in 2^1 ways and the last in 2^0 ways, making for a total of $2^4 2^3 2^2 2^1 2^0 = 2^{10}$ ways altogether, the same number as before. The reason the number of choices goes down by half at every step is that each new quartet already has some of its bases fixed by the earlier choices, leaving bases to be picked only from the complementary pairs that have been left untouched. This point is best appreciated if one carries out the construction on Fig. 10.2 by shading the chosen bases, for one then sees that the new bases that can be picked are the ones that are not in line with the new pure basis picked or one of the ones picked earlier.

The second construction of a decuplet shows that it can be described as a word of five letters, each an E or an O, with the letters from left to right referring to the quartets of the pure bases U through Y that fuse to form it. There are $2^5 = 32$ words altogether, with a word like EOEOO describing not a single decuplet but a whole class of decuplets all of whose members are made up of quartets from the pure bases with the parities indicated.

We now show that only half the 32 words actually describe decuplets. To see this, note that all the decuplets can be obtained from a single one by complementing one or more of its members. However a complementation always causes two of the letters of the word to flip, and so the parity of the word (which we define to be even or odd if the number of O's in it is even or odd) does not change in the process. It therefore follows that all the decuplets must be described by words of a single parity, either even or odd. Which of these is actually the case is easily settled by looking at a single decuplet, and one then finds that the decuplets are described by words of odd parity. Thus words of the type EEEEO, EOEOO or OOOOO are allowed, while those of the type EEEEE, EOEEO or OOEOO are forbidden.

The decuplets fall short of yielding a parity proof for two reasons: firstly, they consist of an even number of bases and, secondly, some of the pure bases contribute odd numbers of each of their rays to the decuplet. However both these defects can be remedied at one stroke by supplementing the decuplet by the pure bases whose quartets contribute O's to its word. Since the number of O's in the word is always odd, one then ends up with a total number of bases that is odd and in which the states of each pure basis occur an even number of times. But this is just a parity proof!

We have therefore arrived at the following rule for extracting a parity proof from Fig. 10.2: pick one member from each pair of complementary hybrids in the outer ring and supplement them by one, three or five pure bases from the inner ring. Because the hybrids in the outer ring (or decuplets) can be picked in $2^{10} = 1024$ ways, there are the same number of parity proofs. One of them, shown in Fig. 10.3, is obtained by picking the decuplet EEEEO in the outer ring and completing it by adding the pure basis Y in the inner ring to get a parity proof described by the word EEEE**E**, with the bold **E** indicating the place where the pure basis has been added to complete the proof.

The number of proofs of the various kinds can be worked out as follows. The decuplets whose words have a single O in them need to have a single pure basis added and so give rise to proofs with 11 bases, while those with three or five O's give rise to proofs with 13 or 15 bases. Any allowed word, such as OEEOO, actually describes 64 different decuplets because the first quartet (letter) can be chosen in 8

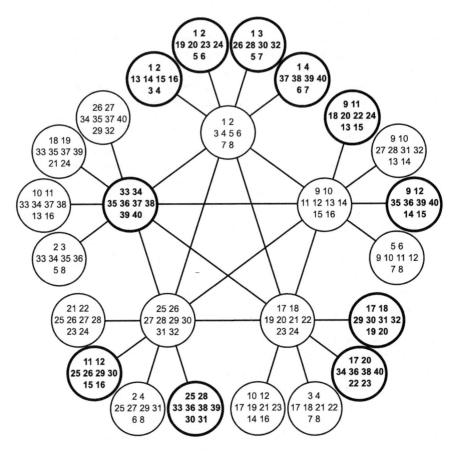

Fig. 10.3 The darkened circles show one of the parity proofs provided by the Kernaghan-Peres states. The ten hybrids in the outer ring, which correspond to the decuplet EEEEO, combine with the pure basis Y, consisting of states 33–40, to give a parity proof described by the word EEEEE whose state-basis count is $8_4 28_2 - 11_8$ (i.e., 8 states occur four times each and 28 states twice each over the 11 bases of the proof). The eight states that occur four times each are 1, 15, 20, 30, 36, 38, 39, 40

ways, the second in 4 ways, the third in 2 ways and the fourth in 1 way,[1] and the product of these numbers is 64. Because the number of words with one, three or five O's is 5, 10 or 1, the number of proofs with 11, 13 or 15 bases can be obtained by multiplying these numbers by 64 to get 320, 640 or 64. The details of these three types of proofs are summarized in Table 3. The Appendix demonstrates that there are no parity proofs apart from these.

---

[1] Recall that each quartet can be chosen in 16 ways, but the restriction to an E or an O cuts the choice down to 8. Also recall that once the first quartet is chosen, the number of possibilities for the second goes down to 4 and that this twofold shrinking continues at each step.

An important feature of these parity proofs is that they are *basis-critical*, by which we mean that they fail if even a single basis is dropped from them. Proofs that are not basis-critical can be whittled down to smaller proofs by shedding bases, and so are redundant from an experimental point of view.

Finally there is the burning question some readers might have: how does the dark line in Fig. 10.1, which is responsible for the magic that takes place there, show up in Fig. 10.2? At first sight there seems to be no telltale sign like it, but a little reflection shows where it is: it lies in the fact that **the decuplets in Fig.10.2 are described by words of odd parity**, for that is what allows them to be promoted into parity proofs.

One can appreciate this point better by stepping back from the particular example of Fig. 10.1 and viewing the problem in a broader context. It turns out that there are many sets of ten 3-qubit observables that can be arranged along the edges of a pentagram in such a way that the observables along each edge commute. Some of these sets have one, three or five dark edges whereas the rest have none (recall that the dark edges are those for which the product of the observables is $-III$ and the light edges those for which it is III, so that only the figures with dark edges give rise to proofs of the KS theorem). If one looks at the Looking Glass equivalent of a figure with only light edges, one sees a diagram very much like Fig. 10.2, consisting of an inner ring of five pure bases and an outer ring of twenty hybrids, *but with the crucial difference that the decuplets obtained from the hybrids are all words of even parity* and so cannot be promoted into parity proofs. The Looking Glass might distort the original but it evidently does so without ruining the magic in it.

We end with a few comments.

The connection between the observables-based KS proof of Fig. 10.1 and its states-based counterpart in Fig. 10.2 is not unique but has been observed in a number of other cases (see Waegell and Aravind 2011 and Table V of Waegell and Aravind 2013, for example). The common pattern in all these cases is that sets of commuting observables yielding an observables-based parity proof give rise to a system of pure and hybrid bases from which a large number of states-based parity proofs can be extracted by picking one member from each complementary pair of hybrids and supplementing them by the needed pure bases; the total number of proofs that can be obtained in all these cases is of the form 2^H, where H is the number of pairs of hybrids involved. The fact that a single observables-based proof gets fragmented into a entire kaleidoscope of states-based proofs seems to be an essential part of the distortion that occurs as one enters the Looking Glass world.

Lisoněk et al. (2014) have shed light on this problem from an alternative viewpoint that makes use of ideas from coding theory. Their approach applies to problems more general than the present one and shows that the total number of parity proofs yielded by a set of bases is always of the form 2^M, where M is an integer related to the kernel of a mapping from the space of the states to the space of the bases. However this estimate includes both the basis-critical proofs as well as many others that can be reduced to them by stripping them of their superfluous bases. The approach of Lisoněk et al provides an interesting counterpoint to the present one and sheds light on aspects of the problem not touched on here. The main goal of the present treatment has been to show how Fig. 10.1 can be taken apart and put back together in the form of Fig. 10.2, and especially to explain how the magic of the dark line in Fig. 10.1 reappears in Fig. 10.2. In other words, it has been to show how one magic trick can be morphed into another.

Parity proofs have by now been discovered in a wide variety of systems (Peres 1991-Lisoněk 2014). We will not attempt to survey them here, but it may be worth pointing out that the Kernaghan-Peres proofs occur many times over among the billions of parity proofs provided by the root vectors of the Lie algebra E8 (Waegell and Aravind 2015).

We hope that this backward look at GHZ, and some of the ways in which it meshes with the work of Mermin, Kernaghan and Peres, would have been of interest to Mike and will miss not being able to get his reaction to it.

Appendix

We show that there are no parity proofs possible with the Kernaghan-Peres states other than the ones listed in Table 10.2. Any parity proof consists of a set of hybrids supplemented by a number of pure bases, if needed. If one looks at the hybrids alone, they must consist of a union of sets of hybrids associated with the various pure bases in such a way that the set associated with any pure basis corresponds to the letter E or O (by which we mean that each state of the pure basis occurs an even or an odd number of times over the hybrids of that set). To determine all the parity proofs that are possible, we must therefore begin by determining all the different ways in which the letters E and O can be picked for each of the pure bases.

Each pure basis has eight hybrids associated with it, namely, the ones at the ends of the lines issuing from it in Fig. 10.2. This set of eight hybrids has $2^8 = 256$ subsets, each consisting of anywhere from zero to eight hybrids. From these subsets we must pick out only those that define an E or O, which are the ones that contain each state

Table 10.2 Parity proofs given by the Kernaghan-Peres states. For each set of ten unpaired hybrids described by words of the type shown in the first column, adding the number of pure bases indicated in the second column leads to a parity proof of the type shown in the third column. The E's in the third column replace the O's in the first column and indicate the pure bases that have been adjoined to the hybrids (or decuplets) of the first column to obtain the parity proof. The fourth column shows the state and basis counts in each of the proofs; for example, the second entry indicates that 14 states occur four times each and 24 states twice each over the 13 bases of the proof. The last column shows the number of proofs of each of these three types, with their sum being $2^{10} = 1024$

Hybrid bases	Pure bases	Parity proof	States-bases	Number of proofs
EEEEO	1	**EEEEE**	$8_4 28_2 - 11_8$	$64 \cdot 5 = 320$
EEOOO	3	**EEEEE**	$14_4 24_2 - 13_8$	$64 \cdot 10 = 640$
OOOOO	5	**EEEEE**	$20_4 20_2 - 15_8$	$64 \cdot 1 = 64$

of the pure basis an even or odd number of times. One finds that there are only 32 subsets meeting this criterion and that they fall into two distinct types that we will call Type 1 and Type 2. Type 1 consists of just the 16 quartets we discussed in the text, with half of them being an E and the other half an O. Type 2 consists of one or more complementary pairs of hybrids from the full set of eight and its members are: one complementary pair (4 members), two complementary pairs (6 members), three complementary pairs (4 members), four complementary pairs (1 member) and the null set (1 member); the total number of possibilities is again 16 and the ones consisting of an even number of pairs are E while the others are O.

To construct a parity proof we must put together a word of five letters, one for each pure basis, and with each letter being an E or an O. We did just this in the text, but restricting the choice of letters for each pure basis to the 16 quartets associated with it. We must now revisit this problem, but allowing the letters for each pure basis to be of both Types 1 and 2. There are now three cases that present themselves for analysis: the letters are all of Type 1 (this is just the problem that was solved in the text); the letters are all of Type 2; and the letters are a mix of Type 1 and Type 2. We will show that the last two cases are ruled out, thus leaving just the solutions found in the text.

Let us begin by disposing of the last case, in which there are a mix of Type 1 and Type 2 letters. Any Type 2 letter involves at least one complementary pair of hybrids and this pair (or pairs) are shared with other pure bases at least one of which has a Type 1 letter associated with it. However the Type 1 letter gets "spoiled" by the intrusion of a pair of complementary hybrids and ceases being a valid letter, causing the attempted proof to fail. Type 1 and Type 2 letters simply cannot coexist, and any attempt to mesh them together to make a parity proof fails.

We are thus left with just the middle case, in which one constructs a five letter word made up of just Type 2 letters. Since each letter corresponds to a set of complementary pairs, the entire word, which is the union of such sets, is itself a set of complementary pairs. The entire universe of words made up of five Type 2 letters consists of all possible subsets of the full set of ten complementary pairs of hybrids. In other words, the "words" we must consider are the sets of $0, 1, 2, \cdots$, or 10 complementary pairs, of which there are $2^{10} = 1024$ altogether. Because each word consists of a certain number of complementary pairs and each pair consists of two hybrids, a word consists of an even number of hybrids. An even number of bases can never give a parity proof, so the word would have to be supplemented by an odd number of pure bases to promote it into a parity proof. The purpose of the added pure bases is to convert all the O's in the word to E's, as required for a parity proof. But, for this process to be successful, it is necessary that there be an odd number of O's in the word so that an odd number of pure bases can be added. However an examination of the 1024 possible words shows that words with an odd number of O's never occur, and so this case can also be dismissed.

References

Bell JS (1964) Physics 1:195
Cabello A, Estebaranz JM, García-Alcaine G (1996) Phys Lett A 212:183
Greenberger DM, Horne MA, Zeilinger A (1989) Bell's Theorem,quantum theory and conceptions of the universe. In: Kafatos M (ed) Kluwer, Dordrecht; Greenberger DM, Horne MA, Shimony A, Zeilinger A (1990) Am J Phys 58:1131
Kernaghan M, Peres A (1995) Phys Lett A 198:1
Lisoněk P, Badziçg P, Portillo JR, Cabello A (2014) Phys Rev A89:042101 (6pp)
Lisoněk P, Raussendorf R, Singh V, Generalized parity proofs of the Kochen-Specker theorem. arXiv:1401.3035v1 [quant-ph]
Mermin ND (1990) Am J Phys 58:731
Mermin ND (1993) Rev Mod Phys 65:803
Pavičić M, Waegell M, Megill ND, Aravind PK (2019) Sci Rep 9:6765
Pavičić M, Megill MD, Merlet JP (2010) Phys Lett A 374:2122; Pavičić M, Merlet JP, McKay BD, Megill ND (2005) J Phys A 38:1577; Pavičić M, Merlet JP, Megill ND (2004) The French National Institute for Research in Computer Science and Control Research Reports RR-5388
Peres A (1991) J Phys A 24:L175
Specker EP (1960) Dialectica 14:239; Kochen S, Specker EP (1967) J Math Mech 17:59. See also Bell JS (1966) Rev Mod Phys 38:447. Reprinted in Bell JS (1987) Speakable and unspeakable in quantum mechanics. Cambridge University Press, Cambridge
Waegell M, Aravind PK (2011) Found Phys 41:1786–1799
Waegell M, Aravind PK (2013) Phys Rev A 88:012102
Waegell M, Aravind PK (2015) J Phys A Math Theor 48:225301
Waegell M, Aravind PK, Megill ND, Pavičić M (2011) Found Phys 41:883